Activities

Fourth Edition

bju press®

Greenville, South Carolina

MATH 5 Student Activities
Fourth Edition

Coordinating Writer
Charlene McCall

Writers
Lindsey Dickinson, MEd
Rita Lovely

Previous Edition Writer
Kathleen Hynicka

Biblical Worldview
Tyler Trometer, MDiv

Academic Oversight
Jeff Heath, EdD
Rachel Santopietro, MEd

Editors
Heather Lonaberger, MA
Abigail Sivyer

Project Coordinator
Kyla J. Smith

Book Designers
Drew Fields
Emily Heinz

Cover Designer
Drew Fields

Page Layout
Maribeth Hayes

Digital Content Management
Peggy Hargis

Illustrators
Lynda Slattery
Courtney Godbey Wise

Cover Illustrators
Craig Oesterling
Del Thompson

Permissions
Kathleen Thompson
Elizabeth Walker

Photo credits: **32** ©iStockphoto.com/James Brey; **82** (bills) BJU Photo Services; **82** (quarters) United States Mint Images; **281** Map Resources

The text for this book is set in Adobe Mathematical Pi, Adobe Minion Pro, Adobe Myriad Pro, Asap and Asap Condensed by Omnibus Type, Helvetica, and Plak by Linotype.

Previously published as *MATH 5 Reviews*

To the Teacher

Math 5 Student Activities provides reviews of core concepts for the student to complete independently. The first page of each two-page lesson reinforces the daily lesson objectives from the *Math 5 Teacher Edition* and provides practice of testable material covered within the corresponding lesson in the *Math 5 Worktext*. This page can also be used as a graded assessment for the lesson. The second page reviews not only concepts from earlier lessons but also concepts mastered in previous grade levels that are used as pre-teaching for upcoming lessons. The Chapter Review pages correspond with the *Math 5 Worktext* Chapter Review pages. These pages provide an excellent study guide for the Chapter Test. The Cumulative Review pages in each chapter provide a spiral review of previous content in a standardized testing format.

Millions Period

Name _____

Write *Ones*, *Thousands*, or *Millions* in the place value chart.

1.

H	T	O ,	H	T	O ,	H	T	O
3	2	9 ,	6	7	5 ,	1	8	7

Write the word for H, T, and O from the place value chart.

2. H _____

3. T _____

4. O _____

Write the number in standard form.

5. _____ = eight million, three hundred twenty-one thousand, six hundred eleven

6. _____ = twenty-nine million, seven hundred forty-two thousand, three hundred thirty-five

7. _____ = 500,000,000 + 20,000,000 + 1,000,000 + 400,000 + 60,000 + 9,000 + 300 + 70 + 8

8. _____ = 1,000,000 + 700,000 + 50,000 + 3,000 + 400 + 90 + 8

9. _____ = (7 × 10,000,000) + (6 × 1,000,000) + (9 × 100,000) + (3 × 10,000) + (1 × 1,000) + (2 × 100) + (4 × 10) + (5 × 1)

10. _____ = (3 × 100,000,000) + (4 × 10,000,000) + (2 × 1,000,000) + (7 × 100,000) + (9 × 10,000) + (6 × 1,000) + (5 × 100) + (8 × 10)

Complete the table.

	Standard form	429,700,000
11.	Word form	
12.	Expanded form	
13.	Expanded form with multiplication	

Write the value of 5 in each number.

14. 152 _____

15. 25,391 _____

16. 51,347,230 _____

17. 147,536 _____

18. 350,248 _____

19. 105,247,369 _____

Write >, <, or = to compare.

20. 3,467 ◯ 3,472   21. 24,057 ◯ 4,989   22. 103,294 ◯ 301,294   23. 1,468 ◯ 10,000 + 8,000

24. 270,016 ◯ two hundred seven thousand, sixteen   25. 81,204 ◯ (8 × 10,000) + (9 × 1,000)

26. 57,943 ◯ 50,000 + 7,000 + 900 + 50 + 2

Write the value.

1.
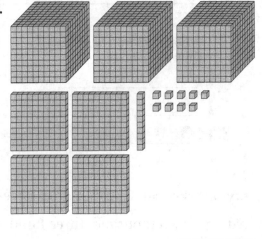

_____

2.
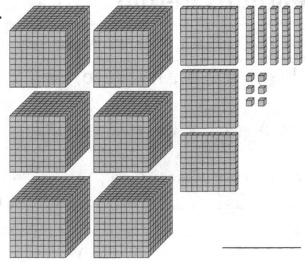

_____

Use the chart to solve the problem.

3. Which islands' populations share the same value in the Tens place?

_____

4. Which islands have a population with a value in the Hundred Thousands place?

_____

5. Would the population of the Cayman and Marshall islands together be greater or less than the population of the Virgin islands?

_____

Islands	Population
Cayman	50,348
Cook	18,723
Marshall	55,449
Virgin	112,940

Put commas in the correct places.

6. 2 9 5 4 6 8   7. 7 2 6 4 1   8. 1 1 1 1 1 1   9. 9 6 7 2 1

Mark the answer.

10. Which number has a 4 in the Hundred Thousands place, a 6 in the Thousands place, and a 0 in the Tens place?

○ 416,203   ○ 406,230   ○ 46,023

11. Which number has a 2 in the Ten Thousands place, a 2 in the Ones place, and a 1 in the Hundreds place?

○ 121,212   ○ 122,112   ○ 121,121

Add.

12. $\begin{array}{r} 17 \\ + 24 \\ \hline \end{array}$
13. $\begin{array}{r} 36 \\ + 18 \\ \hline \end{array}$
14. $\begin{array}{r} 29 \\ + 11 \\ \hline \end{array}$
15. $\begin{array}{r} 54 \\ + 71 \\ \hline \end{array}$
16. $\begin{array}{r} 79 \\ + 47 \\ \hline \end{array}$
17. $\begin{array}{r} 93 \\ + 8 \\ \hline \end{array}$

Subtract.

18. $\begin{array}{r} 91 \\ - 17 \\ \hline \end{array}$
19. $\begin{array}{r} 40 \\ - 29 \\ \hline \end{array}$
20. $\begin{array}{r} 11 \\ - 7 \\ \hline \end{array}$
21. $\begin{array}{r} 23 \\ - 4 \\ \hline \end{array}$
22. $\begin{array}{r} 67 \\ - 18 \\ \hline \end{array}$
23. $\begin{array}{r} 47 \\ - 14 \\ \hline \end{array}$

Billions Period

Write *Ones*, *Thousands*, or *Millions* in the place value chart.

1.

H	T	O ,	H	T	O ,	H	T	O ,	H	T	O
7	4	6 ,	1	2	4 ,	3	5	9 ,	3	8	2

Circle the numbers with a Billions period.

2. 209,437,115   1,070,328,962   45,892,741,301   951,683,297

Match.

_____ 3. 23,345,125

_____ 4. 241,178,000,045

_____ 5. 23,307,981,123

_____ 6. 205,376,148

_____ 7. 249,142,076,389

A the number with the greatest value

B 23 billion, 307 million, 981 thousand, 123

C The value of 5 is 5,000,000.

D twenty-three million, three hundred forty-five thousand, one hundred twenty-five

E 200,000,000,000 + 40,000,000,000 + 1,000,000,000 + 100,000,000 + 70,000,000 + 8,000,000 + 40 + 5

Round the number to the nearest hundred and one thousand.

		hundred	one thousand
8.	3,510		
9.	8,209		
10.	27,498		
11.	74,917		

Round the number to the nearest hundred thousand.

12. 2,937,026 _____

13. 84,592,374 _____

Round the number to the nearest hundred million.

14. 124,509,139 _____

15. 873,209,480 _____

Write the value of the underlined digit.

16. 3,2̲47,890 _____

17. 9̲31,270,105 _____

18. 4̲19,208,319 _____

19. 5̲,137,246,890 _____

Write >, <, or = to compare.

20. 4,278,365 ◯ 4,278,960   21. 32,401,762 ◯ 9,832,407   22. 307,942,860 ◯ 30,794,286

23. 76,000,000,000 ◯ 79 million   24. 2 × 100,000,000 ◯ 199,345,604

Write the number in the place value chart.

H	T	O	,	H	T	O	
Thousands				Ones			
1.			,				forty-six thousand, thirteen
2.			,				seven hundred three thousand, two hundred seventy-four
3.			,				eight hundred sixty-one thousand, one hundred twenty

Compare the numbers. Circle the greater number.

4. 173,206
   173,129

5. 16,455
   16,425

6. 119,206
   19,206

7. 714,413
   714,414

Write >, <, or = to compare.

8. 23,461 $\bigcirc$ 32,416

9. 360,217 $\bigcirc$ 300,000 + 60,000 + 2,000 + 100 + 10 + 7

10. eleven thousand, two hundred twenty-one $\bigcirc$ 11,021

Round each number to the indicated place. Mark the answer.

11. 81,6<u>0</u>5 $\bigcirc$ 81,600
                $\bigcirc$ 81,610

12. 351,<u>2</u>66 $\bigcirc$ 351,200
                $\bigcirc$ 351,300

13. <u>1</u>7,264 $\bigcirc$ 10,000
                $\bigcirc$ 20,000

14. 118,9<u>1</u>9 $\bigcirc$ 118,900
                $\bigcirc$ 118,920

Write the missing number.

15.
```
   22
+  77
_____
```

16.
```
+  77
_____
  110
```

17.
```
   15
+
_____
   76
```

18.
```
   29
+  29
_____
```

19.
```
   63
+   0
_____
```

20.
```
   45
+
_____
   54
```

21.
```
   40
−  17
_____
```

22.
```
   21
−
_____
    9
```

23.
```
−  19
_____
   13
```

24.
```
−  17
_____
   14
```

25.
```
   29
−   6
_____
```

26.
```
   56
−
_____
   11
```

Decimals

Name _____

Write the fraction form and the decimal form that tell what part is shaded.

1.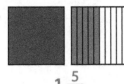
$1\frac{5}{10}$
_____

_____

2.
_____

_____

3.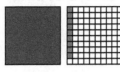
_____

_____

4.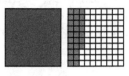
_____

_____

Complete the number line with decimals.

5. 0   1

0   0.1 ____ ____ ____ 0.5 ____ ____ ____ ____ 1.0

6. 3   4

3.0 3.1 ____ ____ ____ ____ ____ ____ ____ ____ 4.0

7. 2.1   2.2

2.10 2.11 2.12 ____ ____ 2.15 ____ ____ 2.18 ____ 2.20

Write the decimal.

8. $8\frac{12}{100}$ = _____

9. $\frac{127}{1,000}$ = _____

10. nine tenths = _____

11. four hundredths = _____

12. $6\frac{6}{10}$ = _____

13. seven tenths = _____

Write the value of the underlined digit.

14. 1.3̲5 _____

15. 4.5̲61 _____

16. 0.817̲ _____

17. 2̲7.13 _____

18. 12̲5.80 _____

19. 38.4̲5 _____

20. 3̲1.5 _____

21. 1.9̲05 _____

Complete the table for 5.27.

22.	**Fraction form**	
23.	**Word form**	
24.	**Expanded form**	
25.	**Expanded form with multiplicaton**	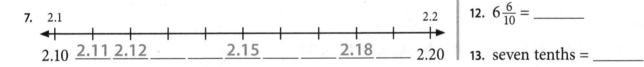 (_____ × _____) + (_____ × _____) + (_____ × _____)

26.
5.2   5.3

5.20 5.21 ____ ____ ____ ____ ____ 5.27 ____ ____ ____ 5.3

Write >, <, or = to compare.

1. 714,213 $\bigcirc$ (7 × 100,000) + (1 × 10,000) + (4 × 1,000) + (2 × 100) + (1 × 10) + (1 × 3)

2. 321,009 $\bigcirc$ thirty-two thousand, nine

3. 69,423 $\bigcirc$ 64,923

Write the value of the underlined digit.

4. 4,218,<u>4</u>06 _____

5. 66,6<u>6</u>1 _____

6. <u>1</u>4,291,423 _____

7. 97,<u>3</u>21,099 _____

8. <u>9</u>,481 _____

9. <u>9</u>16,413,522 _____

Rewrite the number of recycled bottles in standard form and word form.

10. In 2008, volunteers recycled 189,000,000 plastic bottles that littered American public areas.

    standard form: _____

    word form: _____

Match.

_____ 11. 173,263,194

_____ 12. 174,367,497

_____ 13. 172,376,479

_____ 14. 76,734,149

_____ 15. 173,362,194

A The value of 4 is 4,000.

B number with the greatest value

C 173 million + 362 thousand + 194

D one hundred seventy-three million, two hundred sixty-three thousand, one hundred ninety-four

E 100,000,000 + 70,000,000 + 2,000,000 + 300,000 + 70,000 + 6,000 + 400 + 70 + 9

Complete the equation.

16. _____ + 100 = 1,900

17. 230 + _____ = 300

18. 900 − 10 = _____

19. 550 − _____ = 510

20. _____ − 300 = 10

21. 1,500 + _____ = 2,000

22. 420 − 100 = _____

23. _____ − 175 = 0

24. 2,000 + 800 = _____

Equivalent Decimals

Name _____

Mark the equivalent number.

1. 2.1
   - ○ 2.01
   - ○ 21.0
   - ○ 2.10

2. 3
   - ○ 0.3
   - ○ 3.0
   - ○ 0.03

3. 0.15
   - ○ 0.150
   - ○ 0.015
   - ○ 1.500

4. 1.348
   - ○ 13.48
   - ○ 134.8
   - ○ 1.348

5. $\frac{3}{10}$
   - ○ 30
   - ○ 0.3
   - ○ 0.03

6. $\frac{5}{100}$
   - ○ 500
   - ○ 0.005
   - ○ 0.05

7. $2\frac{4}{10}$
   - ○ 2.40
   - ○ 20.4
   - ○ 2.04

8. $\frac{67}{100}$
   - ○ 6.70
   - ○ 0.67
   - ○ 0.067

Write >, <, or = to compare.

9. 3.5 ◯ 3.25

10. 1.783 ◯ 1.789

11. 0.350 ◯ 0.035

12. $1\frac{2}{10}$ ◯ $1\frac{5}{10}$

13. 0.70 ◯ 0.07

14. $3\frac{9}{100}$ ◯ 3.09

Complete the number line with decimals.

15.

16.

Mark the location of the decimal on the number line. Round the decimal to the nearest hundredth.

17. 0.178 rounds to _____.
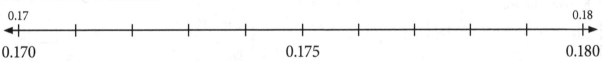

18. 0.324 rounds to _____.
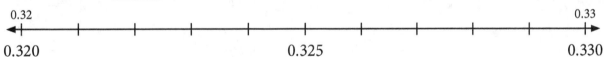

Round the decimal to the nearest one and tenth.

		one	tenth
19.	3.48		3.5
20.	2.07		
21.	12.621		
22.	0.85		

Round to the place with the greatest value. Circle the number.

1.
253,104

    300,000   200,000

2.
5,479,621,086

    5,000,000,000   5,000,000

3.
71,046,843

    700,000   70,000,000

4.
42,671,523,001

    40,000,000,000   40,000,000

Write the value of the underlined digit.

5. 2_1_5,619,433 _____

6. 1,41_6_,746,329 _____

7. _7_27,899 _____

8. 253,4_6_0,612 _____

Count by 10,000s.

9. 36,000 _____ _____ _____ _____

Count by 100,000s.

10. 150,000 _____ _____ _____ _____

Count by 1,000,000s.

11. 28,000,000 _____ _____ _____ _____

Write the temperature.

12.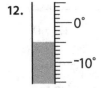

13.

14.

15.

_____   _____   _____   _____

Solve.

16.   413
 + 256

17.   207
 + 437

18.   496
 + 123

19.   738
 + 67

20.   367
 + 167

Positive & Negative Numbers

Name _____

Fill in the missing numbers.

1.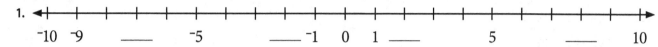
-10 -9 ____ -5 ____ -1 0 1 ____ 5 ____ 10

2.
____ ____ ____ 0 *1* ____ 10

3.
____ ____ ____ 0

Write the number that is marked with a dot.

4. [ ] -6 -5 -4 -3 -2 -1 0 1 2 3 4 5 6

5. [ ] -6 -5 -4 -3 -2 -1 0 1 2 3 4 5 6

6. [ ] -6 -5 -4 -3 -2 -1 0 1 2 3 4 5 6

7. [ ] -6 -5 -4 -3 -2 -1 0 1 2 3 4 5 6

Place a dot on the number line to indicate the number.

8. [8] -3 -2 -1 0 1 2 3 4 5 6 7 8 9

9. [-6] -9 -8 -7 -6 -5 -4 -3 -2 -1 0 1 2 3

10. [-7] -12 -10 -8 -6 -4 -2 0 2 4 6 8 10 12

11. [15] -20 -15 -10 -5 0 5 10 15 20

Write the opposite number.

12. -3 _____   13. 12 _____   14. -34 _____   15. 15 _____   16. 125 _____

Mark the number that shows the increase or decrease.

17.	The elevator went 1 floor below ground level.	○ -1	○ 0	○ 1
18.	The temperature dropped 5 degrees below 0.	○ -5	○ 0	○ 5
19.	The rope climber climbed the twenty-foot rope.	○ -20	○ 0	○ 20
20.	Hannah missed 3 on her spelling test.	○ -3	○ 0	○ 3

Round the number to the nearest one billion, one million, and one thousand.

		one thousand	one million	one billion
1.	17,299,453,501			
2.	609,137,416,022			
3.	3,767,284,963			

Write the number in standard form. Round the number to the nearest one billion.

4. $(6 \times 1{,}000{,}000{,}000) + (5 \times 10{,}000{,}000) + (7 \times 1{,}000{,}000) + (3 \times 100{,}000)$
$+ (2 \times 10{,}000) + (6 \times 1{,}000) + (5 \times 100) + (2 \times 10) + (1 \times 1)$

_____ rounds to _____.
    standard form   nearest billion

5. thirty-two billion, six hundred forty-one million, five hundred twenty-seven thousand, nine hundred sixty-three

_____ rounds to _____.
    standard form   nearest billion

Write the fraction in decimal form.

6. $\dfrac{40}{100} =$ _____

7. $\dfrac{9}{10} =$ _____

8. $\dfrac{78}{100} =$ _____

Write the decimal in fraction form.

9. $0.219 =$ _____

10. $0.67 =$ _____

11. $0.136 =$ _____

Write the decimal.

12. one and four tenths = _____

13. $5\dfrac{6}{10} =$ _____

14. three hundred sixteen thousandths = _____

15. $29\dfrac{111}{1{,}000} =$ _____

Write the missing number.

16.
$$\begin{array}{r} 1{,}417 \\ +\quad 2{,}631 \\ \hline \end{array}$$

17.
$$\begin{array}{r} 4{,}687 \\ +\quad \\ \hline 9{,}251 \end{array}$$

18.
$$\begin{array}{r} 629 \\ +\quad 1{,}737 \\ \hline \end{array}$$

19.
$$\begin{array}{r} 33{,}416 \\ +\quad \\ \hline 53{,}234 \end{array}$$

20.
$$\begin{array}{r} 759 \\ -\quad \\ \hline 324 \end{array}$$

21.
$$\begin{array}{r} \\ -\quad 699 \\ \hline 1{,}343 \end{array}$$

22.
$$\begin{array}{r} 9{,}320 \\ -\quad \\ \hline 5{,}672 \end{array}$$

23.
$$\begin{array}{r} 13{,}471 \\ -\quad 9{,}340 \\ \hline \end{array}$$

Comparing Positive & Negative Numbers

Name _____

Use the word bank to fill in the blanks.
Write the missing numbers.

| positive   negative   increase   decrease |

1. ___negative___ numbers

2. _____ numbers

3.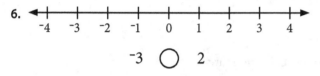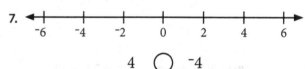

4. Values _____ as you move left.

5. Values _____ as you move right.

Mark the location of the given numbers on the number line. Write > or < to compare.

6.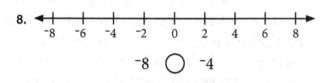

⁻3 ◯ 2

7.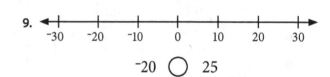

4 ◯ ⁻4

8.
```
 ─┼──┼──┼──┼──┼──┼──┼──┼──
 ⁻8  ⁻6  ⁻4  ⁻2  0   2   4   6   8
```
⁻8 ◯ ⁻4

9.
```
 ─┼──┼──┼──┼──┼──┼──┼──
 ⁻30 ⁻20 ⁻10  0  10  20  30
```
⁻20 ◯ 25

Write the numbers from *least* to *greatest*.

10.
| ⁻4 | 8 | 0 | 5 |

___ ___ ___ ___

11.
| 2 | ⁻2 | 3 | 9 |

___ ___ ___ ___

12.
| 0 | ⁻10 | ⁻20 | 5 |

___ ___ ___ ___

13.
| ⁻8 | 0 | ⁻2 | ⁻5 |

___ ___ ___ ___

14.
| ⁻10 | 4 | 7 | ⁻3 |

___ ___ ___ ___

15.
| 20 | ⁻10 | ⁻5 | 0 |

___ ___ ___ ___

Begin at the dot and follow the jumps. Indicate whether the change is an increase or a decrease.

16.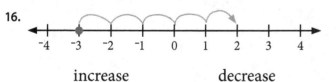

increase   decrease

17.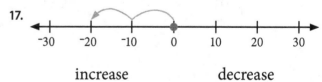

increase   decrease

Write the missing decimals in the blank.

1.
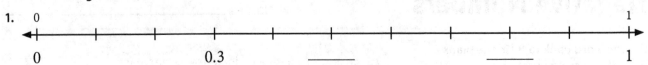

0   0.3   _____   _____   1

2.
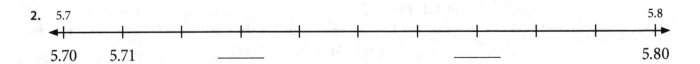

5.70   5.71   _____   _____   5.80

3.
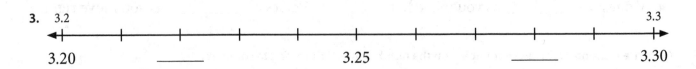

3.20   _____   3.25   _____   3.30

Round the decimal to the nearest one and tenth.

		one	tenth
4.	6.329		
5.	0.7147		
6.	13.906		
7.	1.775		
8.	0.263		
9.	9.490		

Solve. Label your answer.

10. In July 2009, Cape Cod and nearby islands received 6.56 inches of rain. Normal rainfall for the area is 2.78 inches. Did the area receive more or less rain in July 2009 than usual, and how much was the difference?

Write the missing number.

11.
```
    4.67
+
────────
    6.89
```

12.
```
+   5.24
────────
    8.67
```

13.
```
   14.721
+   0.615
─────────
```

14.
```
    0.641
+
─────────
    1.352
```

15.
```
   15.413
−
─────────
    3.101
```

16.
```
    7.077
−
─────────
    7.021
```

17.
```
−   1.243
─────────
    5.534
```

18.
```
   11.389
−   9.041
─────────
```

Number Sense Review

Name _____

Choose a Roman numeral to match the approximate age for each person:
LXXX, XL, XXIII, or X.

1.
_____

2.
_____

3.
_____

4.
_____

Roman Numerals
I = 1
V = 5
X = 10
L = 50
C = 100

Mark the answer. Mark *NH* if the answer is "Not Here."

5.
- ○ 0.03
- ○ 0.3
- ○ 3.00
- ○ NH

6.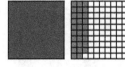
- ○ $1\frac{29}{100}$
- ○ 129
- ○ 1.029
- ○ NH

7.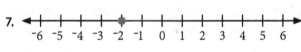
- ○ 2
- ○ 0
- ○ ⁻2
- ○ NH

8.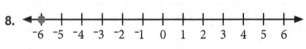
- ○ 1
- ○ ⁻6
- ○ 6
- ○ NH

9. ⁻3 ○ ⁻5
- ○ >
- ○ <
- ○ =

10. 3<u>4</u>.150
- ○ 30.000
- ○ 4.150
- ○ 4.000
- ○ NH

11. 63.2<u>5</u>7
- ○ 0.050
- ○ 0.570
- ○ 50.00
- ○ NH

12. 832,356,754 ○ 83,235,675
- ○ >
- ○ <
- ○ =

13.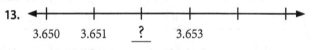
3.650   3.651   ?   3.653
- ○ 3.662
- ○ 3.663
- ○ 3.652
- ○ NH

14. 3 degrees below 0
- ○ ⁻5°
- ○ 5°
- ○ 3°
- ○ NH

Complete the table.

	Standard form	319,426,805
1.	**Word form**	
2.	**Expanded form**	
3.	**Expanded form with multiplication**	

Write the value of the underlined digit.

4. 179,4<u>5</u>3,021 _____

5. <u>3</u>77,129,605,329 _____

6. 2<u>7</u>,436,512,986 _____

7. 4,77<u>3</u>,514 _____

8. <u>3</u>6,947,654 _____

9. 2<u>1</u>8,641,560 _____

Solve. Label your answer.

In 2008, the United States Mint made the last five state quarters. Use the chart to solve the problems.

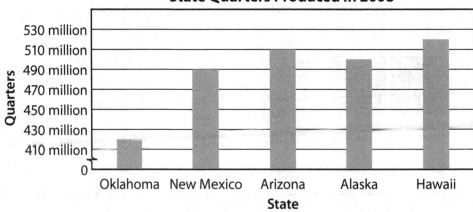

State Quarters Produced in 2008

10. Which state quarter was produced the most in 2008? _____

11. What was the difference between production of Arizona and New Mexico quarters?

_____

12. Write in word form the number of Hawaii quarters produced.

_____

13. Write in expanded form the number of New Mexico quarters produced.

_____

14. What was the total number of Arizona and New Mexico state quarters produced in 2008? _____

Write the sum in word form. _____

Mark the number that is 100 more.

1. 689
   A. 679
   B. 699
   C. 789
   D. 799

2. 463
   A. 363
   B. 453
   C. 473
   D. 563

3. 120
   A. 20
   B. 110
   C. 130
   D. 220

4. 596
   A. 496
   B. 696
   C. 706
   D. 796

5. 1,876
   A. 1,886
   B. 1,976
   C. 2,876
   D. 2,886

6. 4,214
   A. 4,224
   B. 4,314
   C. 4,324
   D. 4,325

Mark the related fact.

7. $7 \times 8 = 56$
   A. $7 \times 7 = 49$ C. $8 \times 7 = 56$
   B. $8 \times 6 = 48$ D. $54 \div 9 = 6$

8. $3 \times 9 = 27$
   A. $27 \div 3 = 9$ C. $3 \times 8 = 24$
   B. $9 \times 2 = 18$ D. $9 \times 4 = 36$

9. $4 \times 6 = 24$
   A. $3 \times 8 = 24$ C. $24 \div 6 = 4$
   B. $24 \div 2 = 12$ D. $24 \div 8 = 3$

10. $40 \div 4 = 10$
    A. $40 \div 5 = 8$ C. $5 \times 8 = 40$
    B. $40 \div 8 = 5$ D. $4 \times 10 = 40$

11. $3 \times 12 = 36$
    A. $12 \times 4 = 48$ C. $2 \times 12 = 24$
    B. $36 \div 3 = 12$ D. $9 \times 4 = 36$

12. $42 \div 7 = 6$
    A. $48 \div 6 = 8$ C. $42 \div 3 = 14$
    B. $8 \times 5 = 40$ D. $6 \times 7 = 42$

Mark the missing addend.

13. ☐ + 6 = 14

  A. 20   C. 8

  B. 10   D. 6

14. 4 + ☐ = 11

  A. 9   C. 7

  B. 8   D. 6

15. 3 + ☐ = 12

  A. 10   C. 7

  B. 9   D. 5

16. ☐ + 8 = 17

  A. 9   C. 6

  B. 7   D. 5

Add.

17.
$$\begin{array}{r} 26 \\ 15 \\ +\,35 \\ \hline \end{array}$$

  A. 57

  B. 66

  C. 76

  D. 87

18.
$$\begin{array}{r} 17 \\ 48 \\ +\,53 \\ \hline \end{array}$$

  A. 108

  B. 118

  C. 129

  D. 138

19.
$$\begin{array}{r} 91 \\ 47 \\ +\,12 \\ \hline \end{array}$$

  A. 130

  B. 139

  C. 140

  D. 150

20.
$$\begin{array}{r} 408 \\ 783 \\ +\,162 \\ \hline \end{array}$$

  A. 1,353

  B. 1,355

  C. 1,360

  D. 1,373

21.
$$\begin{array}{r} 614 \\ 79 \\ +\,46 \\ \hline \end{array}$$

  A. 699

  B. 700

  C. 729

  D. 739

22.
$$\begin{array}{r} 800 \\ 197 \\ +\,213 \\ \hline \end{array}$$

  A. 1,100

  B. 1,200

  C. 1,210

  D. 1,301

Subtract.

23.
$$\begin{array}{r} 986 \\ -\,704 \\ \hline \end{array}$$

  A. 180

  B. 182

  C. 270

  D. 282

24.
$$\begin{array}{r} 580 \\ -\,435 \\ \hline \end{array}$$

  A. 145

  B. 155

  C. 165

  D. 175

25.
$$\begin{array}{r} 600 \\ -\,240 \\ \hline \end{array}$$

  A. 340

  B. 360

  C. 400

  D. 440

Properties

Name _____

Match the equation to the definition.

_____ 1. **Commutative Property of Addition**
The order of addends can be changed without changing the sum.

_____ 2. **Associative Property of Addition**
The grouping of addends can be changed without changing the sum.

_____ 3. **Identity Property of Addition**
When 0 is added to an addend, the sum is that addend.

_____ 4. **Zero Principle of Subtraction**
When 0 is subtracted from a number, the answer is that number.

A $74 + 0 = 74$

B $187 - 0 = 187$

C $(23 + 45) + 98 = 23 + (45 + 98)$

D $159 + 86 = 86 + 159$

Write the value of n that makes the statement true.

5. $n - 0 = 21$

$n =$ _____

6. $35 + n = 35 + 41$

$n =$ _____

7. $28 + n = 28$

$n =$ _____

8. $(4 + 6) + 3 = (4 + n) + 3$

$n =$ _____

9. $92 + n = 92$

$n =$ _____

10. $57 + 45 = n + 57$

$n =$ _____

Complete the table.

11.

Rule: + 5	
10	15
25	
50	
64	

12.

Rule: – 10	
43	33
70	
55	
29	

Solve. Label your answer.

13. Dana made a large insect display. Adrian had 48 insects on her display. Together they collected 106 insects. How many insects were on Dana's display?

Mark the equation that solves the problem.

14. Brent collected 23 insects for his science project. He needs a total of 45 insects to complete the project. How many more insects does Brent need to complete the project?

○ $23 + 45 = 68$ insects
○ $23 + n = 45$ insects
○ $23 + 0 = 45$ insects

Complete the place value chart. Fill in the digits of the expanded number form below.

1. 400,000,000,000 + 30,000,000,000 + 5,000,000,000 + 7,000,000 + 900,000 + 30,000 + 400 + 60 + 7

H	T	O	,	H	T	O	,	H	T	O	,	H	T	O
					Millions									
			,				,				,			

Write the value of the underlined digit.

2. 4,265,843,163 _____

3. 774,955,413 _____

4. 36,568,206,219 _____

5. 704,295,447,308 _____

Write >, <, or = to compare.

6. 214,321,659 ◯ 214,312,695

7. 705,683,217,123 ◯ 705,683,123

8. 123,456,789 ◯ 9,876,543,210

9. 60,521,360 ◯ 60,000,000 + 500,000 + 20,000 + 1,000 + 60

Solve and label.

10. Bright Idea Light Bulb Company sells light bulbs in packages of ten. If they shipped 1,000 packages to the Handy Hardware store, how many light bulbs would be shipped?

Round the number to the nearest one billion, one million, and one thousand.

		one billion	one million	one thousand
11.	7,823,164,508			
12.	29,862,568,743			
13.	418,075,936,214			

Complete the fact.

14.
```
    8
×   9
_____
```

15.
```
    8
×
_____
   56
```

16.
```
×   3
_____
   27
```

17.
```
    6
×
_____
   48
```

18.
```
    6
×
_____
   42
```

19.
```
×   5
_____
   45
```

20.
```
    4
×   8
_____
```

21.
```
×   9
_____
   63
```

22.
```
    9
×   9
_____
```

23.
```
    9
×   4
_____
```

Adding Large Numbers

Name _____

Solve.

1. 4,275
   + 3,912

2. 24,109
   + 50,128

3. 209,417
   + 438,143

4. 63,281
   + 15,749

5. 100,413
   + 298,159

6. 963,210
   + 45,715

7. 559,815
   + 438,345

8. 712,345
   + 98,078

Mark the estimate.

9. The 4th grade has 19 students and the 5th grade has 27 students. About how many paintbrushes will the art teacher need for the 2 classes?
   - ○ 20 paintbrushes
   - ○ 50 paintbrushes
   - ○ 100 paintbrushes

10. The art teacher has 189 craft sticks in a bundle. In another bundle she has 125 craft sticks. About how many craft sticks does she have?
   - ○ 100 craft sticks
   - ○ 200 craft sticks
   - ○ 300 craft sticks

11. In the art closet are 45 tubes of red paint, 53 tubes of yellow paint, and 49 tubes of blue paint. About how many tubes of paint are in the closet?
   - ○ 150 tubes
   - ○ 200 tubes
   - ○ 250 tubes

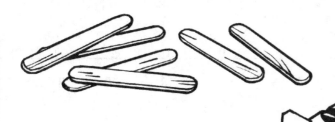

Use the properties to write the missing numbers.

12. $7 + 5 = 5 +$ _____

13. $23 + 0 =$ _____

14. $18 - 0 =$ _____

15. $(4 + 5) + 3 = 4 + (5 +$ _____$)$

16. $47 -$ _____ $= 47$

17. $15 +$ _____ $= 15$

Round to estimate. Solve.

18.
Estimate
2,000
+ 2,000
   1,987
   + 2,304

19.
Estimate
   37,402
   + 15,816

20.
Estimate
   504,917
   + 372,281

Write the decimal.

1. $4\frac{16}{100} =$ _____

2. sixteen and four hundredths = _____

3. $\frac{2}{10} =$ _____

4. four hundred five thousandths = _____

5. $\frac{107}{1,000} =$ _____

6. two thousand, six hundred twenty-two and four tenths = _____

Use the table to find the answer.

Adams Road Baptist Church is making plans for vacation Bible school. They purchased apple juice, grape juice, fruit punch, and bottled water for the children.

Vacation Bible School Drinks		
Drink	Amount Purchased	Left Over
apple juice	6 gallons	2 gallons
grape juice	4 gallons	1 gallon
fruit punch	10 gallons	0 gallons
water	5 gallons	1 gallon

7. How many total gallons of drinks did the church purchase?

   _____

8. How many total gallons were left over? _____

9. Tell the amount of each drink that was used.

   apple juice   grape juice   fruit punch   water

   _____   _____   _____   _____

Complete the number line by writing the decimals.

10.
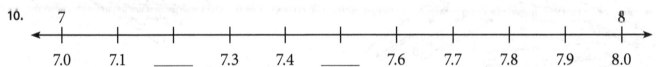

7 | | | | | | | | | | 8
7.0   7.1   ____   7.3   7.4   ____   7.6   7.7   7.8   7.9   8.0

11.
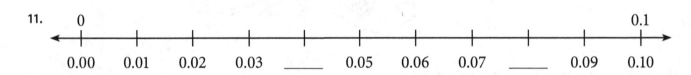

0   0.1
0.00   0.01   0.02   0.03   ____   0.05   0.06   0.07   ____   0.09   0.10

Complete the fact.

12. _?_ × 9 = 54
   ○ 6   ○ 7

13. 7 × 7 = _?_
   ○ 49   ○ 46

14. 9 × 7 = _?_
   ○ 63   ○ 64

15. 7 × 3 = _?_
   ○ 21   ○ 23

16. 4 × 7 = _?_
   ○ 24   ○ 28

17. 5 × _?_ = 30
   ○ 6   ○ 7

18. 7 × 8 = _?_
   ○ 54   ○ 56

19. 6 × 6 = _?_
   ○ 26   ○ 36

Adding Decimals

Name _____

Estimate by rounding to the place of greatest value. Solve.

1. **Estimate**
   30
   + 50

   31.5
   + 45.1

2. **Estimate**

   12.69
   + 29.52

3. **Estimate**

   52.76
   + 85.13

4. **Estimate**

   5.43
   + 1.72

5. **Estimate**

   41.67
   + 29.8

6. **Estimate**

   28.3
   + 6.94

Line up the decimal points. Add.

7. 18 + 2.64 + 37.5 = _____

   Solve
   18.00
   2.64
   + 37.50

8. 27.81 + 10.5 + 34.6 = _____

   Solve

9. 12 + 30.09 + 4.7 = _____

   Solve

10. 1.5 + 32.64 + 8.07 = _____

    Solve

Mark the property that matches the equation.

11. 4 + 5 = 5 + 4

    ○ Identity Property
    ○ Commutative Property

12. 8 + 0 = 8

    ○ Identity Property
    ○ Associative Property

13. (3 + 7) + 4 = 3 + (7 + 4)

    ○ Commutative Property
    ○ Associative Property

Add.

14. $76.45
    + $13.98

15. 203,764
    + 194,305

16. 29.13
    47.6
    + 1.58

17. 7,821
    1,437
    + 1,610

18. $612.43
    $70.94
    + $51.01

Solve.

19. 3,749 + 1,895 = _?_

    Solve

20. 65 + 21.37 = _?_

    Solve

21. 130.4 + 399.8 = _?_

    Solve

Match the equation to the statement that describes it.

_____ **1.** $17 + 39 = 39 + 17$

_____ **2.** $413 + 0 = 413$

_____ **3.** $(18 + 42) + 7 = 18 + (42 + 7)$

_____ **4.** $189 - 0 = 189$

A The addends are grouped differently, but the sum is the same (**Associative Property of Addition**).

B When 0 is subtracted from a number, the answer is that number (**Zero Principle of Subtraction**).

C When 0 is an addend, the sum is the other addend (**Identity Property of Addition**).

D The addends are written in a different order, but the sum is the same (**Commutative Property of Addition**).

Complete the table.

5.

Rule: + 9	
Input	Output
17	
31	
46	
85	

6.

Rule: + 11	
Input	Output
21	
36	
58	
79	

7.

Rule: + 18	
Input	Output
5	
35	
51	
97	

Write the answer.

8. Write an equation that demonstrates the Associative Property of Addition.

_____

9. Write an equation that demonstrates the Commutative Property of Addition.

_____

Write the name of the property for the equation.

10. $7{,}143 + n = 7{,}143$ _____

Circle *True* or *False*.

11. True   False
$$\begin{array}{r} 7 \\ \times\,3 \\ \hline 21 \end{array}$$

12. True   False
$$\begin{array}{r} 4 \\ \times\,7 \\ \hline 28 \end{array}$$

13. True   False
$$\begin{array}{r} 8 \\ \times\,5 \\ \hline 45 \end{array}$$

14. True   False
$$\begin{array}{r} 4 \\ \times\,9 \\ \hline 34 \end{array}$$

15. True   False
$$\begin{array}{r} 5 \\ \times\,5 \\ \hline 25 \end{array}$$

Solve.

16.
$$\begin{array}{r} 71 \\ \times\,\ 3 \\ \hline \end{array}$$

17.
$$\begin{array}{r} 39 \\ \times\,\ 2 \\ \hline \end{array}$$

18.
$$\begin{array}{r} 16 \\ \times\,\ 5 \\ \hline \end{array}$$

19.
$$\begin{array}{r} 47 \\ \times\,\ 3 \\ \hline \end{array}$$

20.
$$\begin{array}{r} 50 \\ \times\,\ 4 \\ \hline \end{array}$$

Subtracting Large Numbers

Name _____

Subtract.

1.
$$\begin{array}{r} \scriptstyle 18 \\ \scriptstyle 2\ 10\quad 6\ 8\ 12 \\ 301{,}792 \\ -\ 180{,}493 \\ \hline \end{array}$$

2.
$$\begin{array}{r} 783{,}210 \\ -\ 592{,}134 \\ \hline \end{array}$$

3.
$$\begin{array}{r} 638{,}019 \\ -\ 134{,}124 \\ \hline \end{array}$$

4.
$$\begin{array}{r} 462{,}109 \\ -\ 331{,}045 \\ \hline \end{array}$$

5.
$$\begin{array}{r} \scriptstyle 9\ \ 9\ 9 \\ \scriptstyle 10\ \ 10\ 10\ 10 \\ 10{,}000 \\ -\ \ 4{,}237 \\ \hline \end{array}$$

6.
$$\begin{array}{r} 107{,}300 \\ -\ \ 23{,}145 \\ \hline \end{array}$$

7.
$$\begin{array}{r} 400{,}000 \\ -\ 132{,}471 \\ \hline \end{array}$$

8.
$$\begin{array}{r} 590{,}000 \\ -\ 321{,}346 \\ \hline \end{array}$$

Round to estimate. Solve.

9.
Estimate
600,000
− 500,000

$$\begin{array}{r} 629{,}178 \\ -\ 493{,}157 \\ \hline \end{array}$$

10.
Estimate

$$\begin{array}{r} 378{,}215 \\ -\ 143{,}102 \\ \hline \end{array}$$

Complete the table.

11.

Rule: + 5	
Input	Output
20	
22	
24	

12.

Rule: + 10	
Input	Output
143	
189	
205	

13.

Rule: − 2	
Input	Output
57	
63	
75	

14.

Rule: − 5	
Input	Output
10	
15	
20	

Round to estimate. Solve.

15.
Estimate

$$\begin{array}{r} 8.33 \\ +\ 3.92 \\ \hline \end{array}$$

16.
Estimate

$$\begin{array}{r} 4{,}167 \\ +\ 5{,}298 \\ \hline \end{array}$$

17.
Estimate

$$\begin{array}{r} 88.30 \\ +\ \ 5.94 \\ \hline \end{array}$$

Use the chart to answer the question.

Worldwide Car Production	
Year	Production
2009	51,971,000
2008	52,941,000
2007	54,920,000
2006	49,887,000

18. What year had the highest car production? _____

19. What is the difference in car production between 2009 and 2008? _____

   Which year had the greater production? _____

20. Arrange the years in order of least production to greatest production. _____

21. The numbers are rounded to what place value? _____

Use the chart to answer the question.

2004 U.S. Apple Production	
Apple Variety	**Units Sold**
Granny Smith	21,966
Gala	25,644
Golden Delicious	31,580
Red Delicious	69,002
Empire	4,994
Fuji	22,396

1. Which variety was sold the most?

   name: _____

   units sold: _____

2. Which variety was sold the least?

   name: _____

   units sold: _____

3. Which 2 varieties had about 22,000 units sold?

   _____ and _____

4. About how many units of Golden Delicious apples were sold? _____

5. Round the units of Empire apples sold to the nearest 100. _____

6. Round the units of Red Delicious apples sold to the nearest 1,000. _____

7. Round the units of Gala apples sold to the nearest 10,000. _____

Complete the chart by listing the apples and their sales in order from *greatest* to *least*.

8.

2004 U.S. Apple Production	
Apple Variety	**Units Sold**

Add.

9. $\begin{array}{r} 54{,}216{,}413 \\ + 26{,}649{,}108 \\ \hline \end{array}$

10. $\begin{array}{r} 107{,}513{,}119 \\ + 341{,}257{,}642 \\ \hline \end{array}$

Write × or ÷ to complete the equation.

11. 54 ◯ 6 = 9   12. 2 ◯ 9 = 18   13. 27 ◯ 3 = 9   14. 64 ◯ 8 = 8

Write >, <, or = to complete the number sentence.

15. 72 ÷ 9 ◯ 32 ÷ 4   16. 4 × 9 ◯ 4 × 8   17. 9 × 7 ◯ 63   18. 7 × 8 ◯ 9 × 6

Write your own equation.

19. _____ × _____ = _____   20. _____ ÷ _____ = _____ ÷ _____

Solve.

21. $\begin{array}{r} 73 \\ \times\ 8 \\ \hline \end{array}$ 22. $\begin{array}{r} 47 \\ \times\ 7 \\ \hline \end{array}$ 23. $\begin{array}{r} 54 \\ \times\ 9 \\ \hline \end{array}$ 24. $\begin{array}{r} 28 \\ \times\ 8 \\ \hline \end{array}$ 25. $\begin{array}{r} 69 \\ \times\ 7 \\ \hline \end{array}$

Subtracting Decimals

Name _____

Match the letter with the mistake in the problem.

1. $32.09 - 14.75 = \underline{\ ?\ }$

$$\begin{array}{r} 32.09 \\ -\ 14.75 \\ \hline 1734 \end{array}$$ _____

2. $85.9 - 1.62 = \underline{\ ?\ }$

$$\begin{array}{r} 85.9 \\ -\ 1.62 \\ \hline 84.32 \end{array}$$ _____

Types of Mistakes

A not lining up decimals

B not annexing zeros

C decimal point not in the answer

3. $79.52 - 3.4 = \underline{\ ?\ }$

$$\begin{array}{r} 79.52 \\ -\ \ 3.4 \\ \hline 791.8 \end{array}$$ _____

4. $4.53 - 1.21 = \underline{\ ?\ }$

$$\begin{array}{r} 4.53 \\ -\ 1.21 \\ \hline 332 \end{array}$$ _____

Solve the problems that are lined up correctly. Place an *X* on problems whose decimals are not in line.

5. $$\begin{array}{r} 13.25 \\ -\ 3.401 \\ \hline \end{array}$$

6. $$\begin{array}{r} 2.31 \\ -\ 1.12 \\ \hline \end{array}$$

7. $$\begin{array}{r} 13.05 \\ -\ \ 6.7 \\ \hline \end{array}$$

8. $$\begin{array}{r} 7.325 \\ -\ 12.13 \\ \hline \end{array}$$

9. $$\begin{array}{r} 13.06 \\ -\ \ 4.13 \\ \hline \end{array}$$

Round to the place of the greatest value to estimate. Solve.

10. $4.125 - 2.031 = \underline{\ ?\ }$

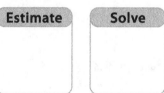

11. $28.03 - 15.4 = \underline{\ ?\ }$

Estimate	Solve

12. $6 - 2.38 = \underline{\ ?\ }$

Estimate	Solve

Match the example to the property.

_____ 13. Associative Property

_____ 14. Commutative Property

_____ 15. Identity Property

_____ 16. Zero Principle of Subtraction

A $8 - 0 = 8$

B $(4 + 3) + 2 = 4 + (3 + 2)$

C $a + b = b + a$

D $7 + 0 = 7$

Use the chart to answer the question.

Top Four 2010 PGA Tour Earnings	
Player	Earnings
1. Steve Stricker	$1,731,000
2. Dustin Johnson	$1,727,450
3. Ian Poulter	$1,442,525
4. Geoff Ogilvy	$1,227,660

1. What are the total earnings of the top 2 golf players? _____

2. What is the difference between the earnings of player 1 and player 4? _____

3. If player 4 doubled his earnings in the next golf tour, about how much money did he earn in that tour? _____

4. Write the earnings of player 3 in expanded form.

_____

Add. Remember to line up the decimal points.

5. $26 + 3.95 + 51.4 =$ _____ | Solve |

6. $70.42 + 16.3 + 11.9 =$ _____ | Solve |

7. $0.9 + 37.23 + 6.19 =$ _____ | Solve |

8. $15 + 87.13 + 7.7 =$ _____ | Solve |

Complete the multiplication chart for the fact families.

9.

×	6	7	8	9
7			56	
8	48			
9				81

Complete the fact.

10. _____ $\times 9 = 27$

11. $4 \times 8 =$ _____

12. $5 \times$ _____ $= 40$

13. $7 \times$ _____ $= 35$

14. _____ $\times 2 = 16$

15. $7 \times$ _____ $= 21$

Mark the correct product.

16. $\begin{array}{r} 42 \\ \times\ 7 \\ \hline ? \end{array}$ ○ 249 ○ 294

17. $\begin{array}{r} 137 \\ \times\ 3 \\ \hline ? \end{array}$ ○ 141 ○ 411

18. $\begin{array}{r} 506 \\ \times\ 5 \\ \hline ? \end{array}$ ○ 2,530 ○ 2,350

19. $\begin{array}{r} 1.18 \\ \times\ 4 \\ \hline ? \end{array}$ ○ 47.2 ○ 4.72

Adding & Subtracting

Name _____

Complete the model. Solve.

1. $53 - n = 25$

$n =$ _____

$n =$ _____

53	
25	

2. $n + 35 = 65$

$n =$ _____

$n =$ _____

65	
35	

3. $21 + n = 45$

$n =$ _____

$n =$ _____

45	
21	

4. $73 - n = 51$

$n =$ _____

$n =$ _____

73	
51	

5. $n - 19 = 60$

$n =$ _____

$n =$ _____

6. $n + 29 = 75$

$n =$ _____

$n =$ _____

Complete the table.

7.

Rule: + 100	
Input	Output
145	**245**
307	
295	

8.

Rule: − 50	
Input	Output
75	
100	
250	

9.

Rule: + 25	
Input	Output
10	
25	
100	

10.

Rule: − 10	
Input	Output
80	
100	
125	

Round to estimate. Solve.

11. Estimate

$\begin{array}{r} 32{,}179 \\ + 49{,}384 \end{array}$

12. Estimate

$\begin{array}{r} 8{,}193 \\ - 2{,}041 \end{array}$

13. Estimate

$\begin{array}{r} 56{,}104 \\ - 31{,}005 \end{array}$

Solve. Label your answer.

14. Luke bought a basketball that cost $14.98. How much change did he receive from $20.00?

15. Luke priced new basketball shoes at $59.99 and track shoes at $70.00. What is the difference in the cost of the shoes?

16. Luke has 2,489 cards in his basketball card collection. His goal is to have 5,000 in his collection. How many more cards does he need to meet his goal?

17. Luke's friend has 12,008 cards. Compare the number of cards in the collections of Luke and his friend.

© 2021 BJU Press. Reproduction prohibited.

Use the data from the chart to find the answer.

Picnic Menu and Prices	
Supplies	Prices
chicken fingers	$22.79
fruit salad	$13.46
drinks	$12.00
vegetables and dip	$9.45

1. Mrs. Peterson's class collected money for a fall picnic. They were able to collect $60.00. Will they have enough money to pay for the food that they plan to use for the picnic? Explain.

_____

_____

_____

2. What were the most and least expensive items for the picnic?

most expensive item: _____   price: _____

least expensive item: _____   price: _____

3. Is the total price of fruit salad and vegetables and dip more or less than the price of the chicken fingers? _____

Write a number sentence to explain the difference.

_____

4. Mrs. Peterson bought her class ice cream as a special treat. She paid twenty extra dollars for the ice cream. What was the total expense of the picnic?

_____

Subtract.

5.   129,463,105
 −   4,312,301

6.   411,563,478
 − 268,315,410

7.   916,433
 − 63,321

8.   189,465,362
 − 89,364,653

Complete the multiplication chart.

	×	5	6	7	8	9	10	11	12
9.	5		30		40				
10.	6	30			48				
11.	7			49					
12.	8	40	48			72			
13.	9				72				

Addition & Subtraction Review

Name _____

Mark the answer.

1. $141 + 267 = 267 + n$

   $n = \underline{\ ?\ }$

   ○ 408
   ○ 141
   ○ 126
   ○ 0

2. $a - 0 = \underline{\ ?\ }$

   ○ a
   ○ b
   ○ c
   ○ n

3.
   $$
   \begin{array}{r}
   248,765 \\
   + 103,271 \\
   \hline
   352,?36
   \end{array}
   $$

   ○ 3
   ○ 2
   ○ 1
   ○ 0

4. $428,391 - 201,140 = \underline{\ ?\ }$

   ○ 227,440
   ○ 227,231
   ○ 227,251
   ○ 228,250

5. $5.83 + 1.109 = \underline{\ ?\ }$

   ○ 16.92
   ○ 1.692
   ○ 5.93
   ○ 6.939

6. $8 - 3.25 = \underline{\ ?\ }$

   ○ 5.25
   ○ 4.75
   ○ 3.23
   ○ 5.75

Use the chart to mark the answer.

Population & Land Area		
State	Population	Square Miles
Alaska	710,249	663,267
California	37,252,895	163,696
Florida	18,804,623	65,755
New York	19,378,087	54,556
Texas	25,146,105	268,581

7. Which state has the largest land area?

   ○ Alaska   ○ Florida
   ○ California   ○ Texas

8. Which state has the greatest population?

   ○ Alaska   ○ Florida
   ○ California   ○ Texas

9. Which state has about 700,000 people?

   ○ Alaska   ○ Florida
   ○ California   ○ Texas

10. Which state has about 50,000 square miles?

    ○ Alaska   ○ New York
    ○ Florida   ○ Texas

11. Which states have over 20,000,000 people? (Mark two.)

    ○ Alaska   ○ New York
    ○ California   ○ Texas

12. Which state is the least populated?

    ○ Alaska   ○ New York
    ○ California   ○ Texas

Mark the answer.

13.

57	
32	n

$n = \underline{\ ?\ }$

○ 5 × 32
○ 32 + 57
○ 30
○ 25

14.
$$\begin{array}{r} 23{,}841 \\ + 59{,}320 \\ \hline ? \end{array}$$

○ 83,161
○ 79,298
○ 83,160
○ 83,000

15. $7.3 + 5.9 = \underline{\ ?\ }$

○ 10.5
○ 13.9
○ 12.9
○ 13.2

16. $81 + 94 = \underline{\ ?\ }$

○ 175
○ 174
○ 173
○ 172

17.

Rule: + 5	
Input	**Output**
37	?

○ 40
○ 41
○ 42
○ 43

18.

Rule: − 12	
Input	**Output**
48	?

○ 12
○ 24
○ 36
○ 40

19.

n	
45	25

$n = \underline{\ ?\ }$

○ 55
○ 60
○ 65
○ 70

20.
$$\begin{array}{r} 1{,}000 \\ - \quad 296 \\ \hline ? \end{array}$$

○ 700
○ 704
○ 896
○ 804

21. $29 - 0 = 29$

○ Associative Property
○ Commutative Property
○ Identity Property
○ Zero Principle

22.
$$\begin{array}{r} 241 \\ 735 \\ + 869 \\ \hline ? \end{array}$$

○ 1,745
○ 1,846
○ 1,845
○ 2,845

23.
$$\begin{array}{r} 5.63 \\ - 1.28 \\ \hline ? \end{array}$$

○ 4.35
○ 4.05
○ 4.45
○ 5.00

24. $34.1 + 27.95 = \underline{\ ?\ }$

○ 31.36
○ 62.05
○ 621.3
○ 6,213

Mark the period name for the place with the greatest value.

1. 16,073,982

   A. Ones   C. Millions
   B. Thousands   D. Billions

2. 5,387,982,421

   A. Ones   C. Millions
   B. Thousands   D. Billions

3. 960,000

   A. Ones   C. Millions
   B. Thousands   D. Billions

4. 178,594,421,620

   A. Ones   C. Millions
   B. Thousands   D. Billions

Round the number to the nearest hundred thousand.

5. 65,372,608

   A. 6,000,000   C. 65,300,000
   B. 60,000,000   D. 65,400,000

6. 423,786

   A. 4,000,000   C. 500,000
   B. 400,000   D. 420,000

7. 1,506,298

   A. 1,500,000   C. 1,600,000
   B. 1,510,000   D. 2,000,000

Use the given numbers to find the answer.

| A. 178,640,596 | C. 17,830,641,592 |
| B. 1,178,640,596 | D. 7,830,596 |

8. the largest number

   A   C
   B   D

9. The value of 8 is 800,000,000.

   A   C
   B   D

10. The value of 7 is 7,000,000.

    A   C
    B   D

11. 178 million, 640 thousand, 596

    A   C
    B   D

12. The greatest place value is one billion.

    A   C
    B   D

13. the number with the least value

    A   C
    B   D

14. 17 billion, 830 million, 641 thousand, 592

    A   C
    B   D

The numbers are ordered *least* to *greatest*. Mark the missing number.

Mark the number line that shows the correct location of the given number.

15.

7,381,460
7,592,731
?
9,577,730

A. 7,459,273
B. 7,582,731
C. 9,507,302
D. 9,587,375

16.

?
5,903,618
5,930,618
5,933,618

A. 5,913,618
B. 5,903,000
C. 5,930,800
D. 6,000,000

17.

21,268,195
?
21,342,180
23,650,981

A. 20,583,290
B. 21,114,617
C. 21,300,416
D. 22,103,483

18.

498,217
498,526
498,851
?

A. 499,900
B. 498,741
C. 498,622
D. 498,123

19.

11.132
?
12.713
14.067

A. 10.175
B. 11.099
C. 11.102
D. 12.052

20. 48

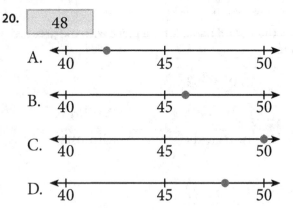

21. 795

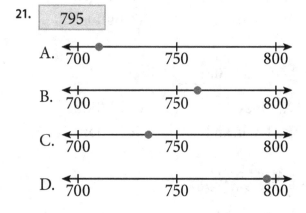

22. 3,361

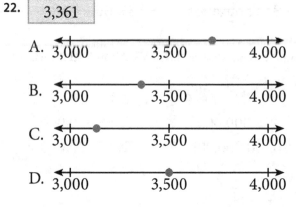

23. ⁻6

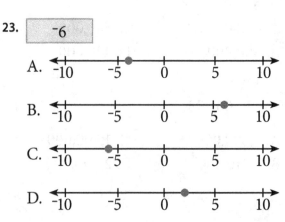

Multiplication Properties

Match the property with the example.

_____ 1. $1 \times 58 = 58$

_____ 2. $21 \times 45 = 45 \times 21$

_____ 3. $76 \times 0 = 0$

_____ 4. $(5 \times 34) \times 12 = 5 \times (34 \times 12)$

Multiplication Properties

A Associative Property

B Identity Property

C Commutative Property

D Zero Property

Match the numerical expression with the phrase.

_____ 5. three sets of six

_____ 6. twelve and eight more

_____ 7. two less than 17

_____ 8. four groups of eight

A 4×8

B $17 - 2$

C 3×6

D $12 + 8$

Solve.

9. $\begin{array}{r} 3 \\ \times 5 \\ \hline \end{array}$
10. $\begin{array}{r} 7 \\ \times 2 \\ \hline \end{array}$
11. $\begin{array}{r} 5 \\ \times 8 \\ \hline \end{array}$
12. $\begin{array}{r} 4 \\ \times 9 \\ \hline \end{array}$
13. $\begin{array}{r} 6 \\ \times 1 \\ \hline \end{array}$
14. $\begin{array}{r} 7 \\ \times 4 \\ \hline \end{array}$
15. $\begin{array}{r} 8 \\ \times 0 \\ \hline \end{array}$
16. $\begin{array}{r} 3 \\ \times 9 \\ \hline \end{array}$

17. $9 \times 8 =$ _____   18. $7 \times 6 =$ _____   19. $8 \times 4 =$ _____   20. $6 \times 6 =$ _____

Write a multiplication equation for each picture.

21.

22.

23.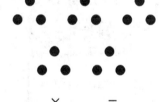

_____ × _____ = _____   _____ × _____ = _____   _____ × _____ = _____

Multiply.

24. $6 \cdot 3 =$ _____   25. $7 \cdot 5 =$ _____   26. $8 \cdot 9 =$ _____   27. $9 \cdot 7 =$ _____

Write a multiplication equation for each addition equation.

28. $6 + 6 =$ _____

_____ × _____ = _____

29. $8 + 8 + 8 + 8 =$ _____

_____ × _____ = _____

30. $7 + 7 + 7 =$ _____

_____ × _____ = _____

31. $1 + 1 + 1 + 1 + 1 =$ _____

_____ × _____ = _____

Use the data from the graph to find the answer and complete the chart.

Average Daily Milk Sales

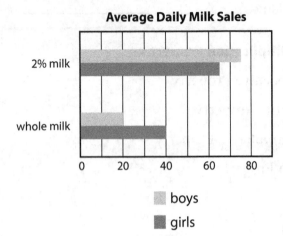

1.

	2% milk	whole milk	2% and whole milk
boys		20	
girls	65		
total students	140	60	

Match the property used to solve the problem.

Multiplication Properties

A Associative Property

B Identity Property

C Commutative Property

D Zero Property

_____ **2.** $295 \times 3 = 3 \times 295$
Reorder factors.

_____ **3.** $17 \times 0 = 0$
0 is a factor.

_____ **4.** $59 \times 1 = 59$
1 is a factor.

_____ **5.** $4 \times (5 \times 307) = (4 \times 5) \times 307$
Regroup factors.

Check the bags to determine whether the correct change was given.
Circle *Y* for yes and *N* for no.

6.
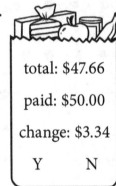
total: $47.66

paid: $50.00

change: $3.34

Y   N

7.
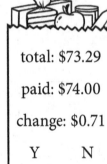
total: $73.29

paid: $74.00

change: $0.71

Y   N

8.
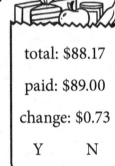
total: $88.17

paid: $89.00

change: $0.73

Y   N

9.
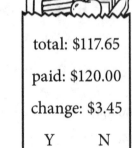
total: $117.65

paid: $120.00

change: $3.45

Y   N

Solve.

10. $6 \times$ _____ $= 42$

11. $5 \times$ _____ $= 40$

12. $9 \times$ _____ $= 27$

13. $7 \times 4 =$ _____

14. _____ $\times 9 = 72$

15. $3 \times$ _____ $= 24$

16. _____ $\times 7 = 35$

17. $2 \times$ _____ $= 16$

18. _____ $\times 7 = 49$

19. $8 \times 7 =$ _____

20. $8 \times 3 =$ _____

21. _____ $\times 4 = 36$

Prime & Composite Numbers

Name _____

Write the multiples of the number.

1. 8 ____ ____ ____ ____ ____ ____ ____ ____ ____
       1×8 2×8 3×8 4×8 5×8 6×8 7×8 8×8 9×8

2. 4 ____ ____ ____ ____ ____ ____ ____ ____ ____
       1×4 2×4 3×4 4×4 5×4 6×4 7×4 8×4 9×4

3. 3 ____ ____ ____ ____ ____ ____ ____ ____ ____

Circle the multiples that are common to each set. List the common multiples.

4. 2   2   4   ⑥   8   10   ⑫
    6   ⑥   ⑫   18   24   30   36   __6, 12__
                                                   common multiples

5. 3   3   6   9   12   15   18
    6   6   12   18   24   30   36   _____
                                                   common multiples

6. 5   5   10   15   20   25   30
    10   10   20   30   40   50   60   _____
                                                   common multiples

Write the factors of each number. Write the factors in order from *least* to *greatest*.

7. 4   8. 7   9. 5   10. 9

    1 × ____   ____ × ____   ____ × ____   ____ × ____
    2 × ____   ____ , ____   ____ , ____   ____ × ____
    1 , ____ , ____   ____ , ____ , ____

Label the number *prime* or *composite*.

11. 3 1×3 _____   12. 9 $1 \times 9, 3 \times 3$ _____

13. 6 $1 \times 6, 2 \times 3$ _____   14. 5 1×5 _____

15. 7 1×7 _____   16. 4 $1 \times 4, 2 \times 2$ _____

Write the number in expanded form.

1. 4.136 = $\underline{4 + 0.1 + 0.03 + 0.006}$

2. 7.031 = _____

3. 20.18 = _____

4. 5.633 = _____

5. 1.259 = _____

6. 0.43 = _____

7. 27.68 = _____

8. 1.69 = _____

9. 911.4 = _____

Write the value of the underlined digit.

10. 819.45<u>2</u> _____

11. 4<u>7</u>.03 _____

12. 5.<u>9</u>18 _____

13. 21.7<u>4</u>1 _____

Write >, <, or = to compare.

14. 413.214 ◯ 413.20

15. 9.425 ◯ 9.452

16. ⁻3,641 ◯ 3,641

17. 27.48 ◯ 27.480

Write the opposite number.

18. ⁻5 _____

19. 216 _____

20. 4,579 _____

21. 37 _____

Write the number marked with a dot on the number line.

22. _____

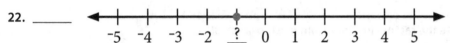

23. _____

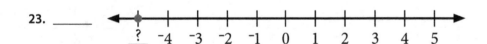

24. _____

Solve.

25.
$\begin{array}{r} 19 \\ \times\ \ 4 \\ \hline \end{array}$

26.
$\begin{array}{r} \ \ \ \\ \times\ \ 2 \\ \hline 412 \end{array}$

27.
$\begin{array}{r} 3.6 \\ \times\ \ 4 \\ \hline \end{array}$

28.
$\begin{array}{r} 17 \\ \times\ \ \ \\ \hline 51 \end{array}$

29.
$\begin{array}{r} 65 \\ \times\ \ 3 \\ \hline \end{array}$

30.
$\begin{array}{r} 4.4 \\ \times\ \ \ \\ \hline 13.2 \end{array}$

Distributive Property

Name _____

Write a multiplication problem for each picture. Solve.

1.

$$4 \times \underline{\hspace{1cm}} = \underline{\hspace{1cm}}$$

2.

$$5 \times \underline{\hspace{1cm}} = \underline{\hspace{1cm}}$$

Use the Distributive Property to solve.

3.

$$3 \times 41 = \underline{\quad ? \quad}$$

$$3 \times (\underline{\ 40\ } + \underline{\ 1\ }) =$$

$$(\underline{\ 3\ } \times \underline{\ 40\ }) + (\underline{\ 3\ } \times \underline{\ 1\ }) =$$

$$\underline{\hspace{1cm}} + \underline{\hspace{1cm}} = \underline{\hspace{1cm}}$$

4.

$$4 \times 26 = \underline{\quad ? \quad}$$

$$4 \times (\underline{\ 20\ } + \underline{\hspace{1cm}}) =$$

$$(4 \times \underline{\hspace{1cm}}) + (4 \times \underline{\hspace{1cm}}) =$$

$$\underline{\hspace{1cm}} + \underline{\hspace{1cm}} = \underline{\hspace{1cm}}$$

5.

$$2 \times 53 = \underline{\quad ? \quad}$$

$$2 \times (\underline{\hspace{1cm}} + \underline{\hspace{1cm}}) =$$

$$(\underline{\hspace{1cm}} \times \underline{\hspace{1cm}}) + (\underline{\hspace{1cm}} \times \underline{\hspace{1cm}}) =$$

$$\underline{\hspace{1cm}} + \underline{\hspace{1cm}} = \underline{\hspace{1cm}}$$

6.

$$6 \times 34 = \underline{\quad ? \quad}$$

$$6 \times (\underline{\hspace{1cm}} + \underline{\hspace{1cm}}) =$$

$$(6 \times \underline{\hspace{1cm}}) + (6 \times \underline{\hspace{1cm}}) =$$

$$\underline{\hspace{1cm}} + \underline{\hspace{1cm}} = \underline{\hspace{1cm}}$$

Use mental math to solve.

7. $4 \times 200 = \underline{\hspace{2cm}}$

8. $3 \times 60 = \underline{\hspace{2cm}}$

9. $50 \times 4 = \underline{\hspace{2cm}}$

10. $7 \times 80 = \underline{\hspace{2cm}}$

11. $1{,}000 \times 9 = \underline{\hspace{2cm}}$

12. $100 \times 5 = \underline{\hspace{2cm}}$

Write the multiples.

13. $\boxed{3}$ _____ _____ _____ _____ _____

14. $\boxed{5}$ _____ _____ _____ _____ _____

Write the factors of the number from *least* to *greatest*.

15. $\boxed{6}$ $\underline{1 \times 6,\ 2 \times 3}$ _____ _____

16. $\boxed{9}$ $\underline{1 \times 9,\ 3 \times 3}$ _____ _____

Write the factors. Label the number *prime* or *composite*.

17. $\boxed{3}$ $\underline{\quad 1,\ 3 \quad}$ $\underline{\quad prime \quad}$

18. $\boxed{8}$ _____ _____

19. $\boxed{10}$ _____ _____

20. $\boxed{7}$ _____ _____

Think of the property to complete the equation.

_____ 1. $0 \times 8 = \underline{}$ **A** $(c \cdot d) \cdot e$

_____ 2. $(3 \times 6) \times 4 = \underline{}$ **B** $3 \times (6 \times 4)$

_____ 3. $5 \times 1 = \underline{}$ **C** 8×0

_____ 4. $c \cdot (d \cdot e) = \underline{}$ **D** 1×5

_____ 5. $b \cdot (c \cdot d) = \underline{}$ **A** 7×0

_____ 6. $(1 \times 2) \times 3 = \underline{}$ **B** 3×4

_____ 7. $0 \times 7 = \underline{}$ **C** $1 \times (2 \times 3)$

_____ 8. $4 \times 3 = \underline{}$ **D** $(b \cdot c) \cdot d$

Write a mathematical expression for the phrase.

9. three sets of eight ___ 3×8 ___

10. four and two more _____

11. seven added to zero _____

12. eleven minus two _____

13. five less than twelve _____

14. nine in each of four sets _____

Use the information from the graph to find the answer.

15. Faith Christian School has a special celebration for summer birthdays. What is the total for June–August?

16. One month has twice as many birthdays as another. List the two months and their totals.

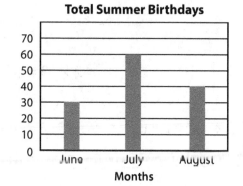

Total Summer Birthdays

Complete the table.

17.

Rule: × 2	
Input	**Output**
13	
25	
34	
47	

18.

Rule: × 3	
Input	**Output**
17	
31	
44	
52	

19.

Rule: × 5	
Input	**Output**
9	
16	
18	
21	

1-Digit Multipliers

Name _____

Mark the equation that matches the array.

1.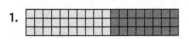

   ○ $10 \times (3 + 8) = n$
   ○ $3 \times (10 + 8) = n$

2.

   ○ $10 + (10 + 1) = n$
   ○ $(5 \times 20) + (5 \times 1) = n$

3.

   ○ $4 \times (10 + 9) = n$
   ○ $100 + (90 + 4) = n$

Use the Distributive Property to solve.

4. $9 \times 24 = n$

   $\underline{\ 9\ } \times (\underline{\ 20\ } + \underline{\quad}) =$

   $(\underline{\quad} \times \underline{\quad}) + (\underline{\quad} \times \underline{\quad}) =$

   $\underline{\quad} + \underline{\quad} =$

   $\underline{\quad} = n$

5. $7 \times 43 = n$

   $\underline{\quad} \times (\underline{\quad} + \underline{\quad}) =$

   $(\underline{\quad} \times \underline{\quad}) + (\underline{\quad} \times \underline{\quad}) =$

   $\underline{\quad} + \underline{\quad} =$

   $\underline{\quad} = n$

Mark the estimate.

6. $\begin{array}{r} 49 \\ \times\ 3 \\ \hline ? \end{array}$ ○ 120 ○ 150

7. $\begin{array}{r} 724 \\ \times\ 4 \\ \hline ? \end{array}$ ○ 2,800 ○ 3,200

8. $\begin{array}{r} 513 \\ \times\ 8 \\ \hline ? \end{array}$ ○ 4,000 ○ 4,800

9. $\begin{array}{r} 3,806 \\ \times\ 7 \\ \hline ? \end{array}$ ○ 21,000 ○ 28,000

Solve.

10. $\begin{array}{r} 76 \\ \times\ 2 \\ \hline \end{array}$

11. $\begin{array}{r} 1,062 \\ \times\ 5 \\ \hline \end{array}$

12. $\begin{array}{r} \$2.35 \\ \times\ 9 \\ \hline \end{array}$

13. $\begin{array}{r} 241 \\ \times\ 6 \\ \hline \end{array}$

14. $\begin{array}{r} 359 \\ \times\ 4 \\ \hline \end{array}$

Label the number *prime* or *composite*.

15. ☐ 5 _____

16. ☐ 10 _____

17. ☐ 8 _____

18. ☐ 11 _____

Write the multiples.

19. ☐ 4 ___ ___ ___ ___ ___ ___ ___ ___

20. ☐ 6 ___ ___ ___ ___ ___ ___ ___ ___

Write the factors from *least* to *greatest*.

21. ☐ 9 ___ ___ ___

22. ☐ 10 ___ ___ ___ ___

23. ☐ 6 ___ ___ ___ ___

24. ☐ 12 ___ ___ ___ ___ ___ ___

Write the factors of the number from *least* to *greatest*. Label the number *prime* or *composite*.

1. 6 _____ _____

2. 17 _____ _____

3. 12 _____ _____

4. 45 _____ _____

5. 21 _____ _____

6. 54 _____ _____

Use the chart to find the answer.

Population in 2000	
Elkhart, IN	51,874
Oshkosh, WI	62,916
Kent, WA	79,524
Charleston, SC	96,650
Sandy, UT	88,418
Loveland, CO	50,608

7. Add the populations of the three largest cities. Mark the correct sum.

   ○ 249,624   ○ 26,592
   ○ 264,592   ○ 20,114

8. Add the populations of the last two cities on the chart. Write the sum.

   _____

9. Subtract the smallest population from the largest population. Write the difference.

   _____

10. Which two cities' populations round to 50,000? What is the difference between them?

   _____

Solve.

11. $5\overline{)25}$

12. $49 \div 7 =$ _____

13. $23\overline{)46}$

14. $121 \div 11 =$ _____

15. $81 \div 9 =$ _____

16. $6\overline{)54}$

17. $\frac{40}{8} =$ _____

18. $3\overline{)69}$

19. $\frac{35}{7} =$ _____

2-Digit Multipliers

Name _____

Round to estimate. Solve.

1. [Estimate]   31
                 × 48

2. [Estimate]   289
                 × 13

3. [Estimate]   423
                 × 61

Solve.

4.   84
    × 67

5.   380
    × 15

6.   508
    × 32

7. $1.45
    ×   12

Write the multiples.

8. [2] _____ 4 _____ _____ _____ _____ _____ _____ _____ _____ _____

9. [10] _____ _____ 30 _____ _____ _____ _____ _____ _____ _____

Use the Distributive Property to solve.

10.   _____ × (90 + 8) = n

    (_____ × _____) + (_____ × _____) = n

                    _____ + _____ = n

                            _____ = n

11. $7 × 83 = n$

    _____

    _____

    _____

    _____

Use the information in the chart to answer the questions.

Fast-Food Calorie Count	
burger	540
chicken bites	380
chicken wrap	185
grilled chicken	350

12. Which fast food choice has fewer calories: burger or chicken?

    _____

13. Which has more calories: a burger or 2 chicken wraps?

    _____

14. Which has more calories: chicken bites or 2 chicken wraps?

    _____

15. Which 2 choices have about the same number of calories?

    _____

Use the Distributive Property to solve.

1.

$4 \times 32 = \underline{\ ?\ }$

$4 \times (30 + 2) =$

$(4 \times \underline{\quad}) + (4 \times \underline{\quad}) =$

$\underline{\quad} + \underline{\quad} = \underline{\quad}$

2.

$5 \times 26 = \underline{\ ?\ }$

$5 \times (\underline{\quad} + \underline{\quad}) =$

$(5 \times \underline{\quad}) + (5 \times \underline{\quad}) =$

$\underline{\quad} + \underline{\quad} = \underline{\quad}$

Write *prime* if the number has no other factors.
Write another expression if the number is composite.

3. 16

1×16

_____

4. 3

1×3

_____

5. 12

1×12

_____

6. 25

1×25

_____

7. 17

1×17

_____

Write the factors of the numbers from *least* to *greatest*. Label the number as *prime* or *composite*.

8. 39 _____ _____

9. 23 _____ _____

10. 27 _____ _____

11. 15 _____ _____

Complete the chart.

12.

H	T	O	,	H	T	O	,	H	T	O	,	H	T	O
2	1	3	,	5	6	0	,	1	8	7	,	2	6	4

Complete the table.

13.	**Standard form**	213,_____,_____,_____
14.	**Word form**	
15.	**Expanded form**	200,000,000,000 + _____ + 3,000,000,000 + _____ + _____ + 100,000 + _____ + _____ + 200 + _____ + 4

Mark the correct answer.

16. $20 \times 30 =$
   ○ 60
   ○ 600
   ○ 6,000

17. $200 \times 50 =$
   ○ 100
   ○ 1,000
   ○ 10,000

18. $70 \times 300 =$
   ○ 2,100
   ○ 21,000
   ○ 210,000

Multiplying & Estimating

Name _____

Round to estimate. Solve.

1. | Estimate |

$$\begin{array}{r} 4\ 1\ 3 \\ \times\quad 2\ 7 \\ \hline \end{array}$$

2. | Estimate |

$$\begin{array}{r} 6\ 9\ 5 \\ \times\quad 4\ 8 \\ \hline \end{array}$$

Solve.

3. $\begin{array}{r} \$75.31 \\ \times\quad 19 \\ \hline \end{array}$

4. $\begin{array}{r} 5{,}102 \\ \times\quad 56 \\ \hline \end{array}$

5. $\begin{array}{r} 8{,}320 \\ \times\quad 24 \\ \hline \end{array}$

6. $\begin{array}{r} 2{,}714 \\ \times\quad 35 \\ \hline \end{array}$

Complete the table.

7.

a	$a \cdot 5$
8	40
10	
12	

8.

a	$a \cdot 12$
7	
15	
25	

9.

a	$a \cdot 25$
2	
4	
10	

Write the multiples of the numbers.

10. | 2 | 2   4   ____ ____ ____ ____ ____ ____

     | 4 | 4   ____ ____ ____ ____ ____ ____

List the common multiples of 2 and 4.

11. _____, _____, _____, _____

Use mental math to solve. Complete the equation.

12. $4 \times 200 =$ _____

13. $3 \times 1{,}000 =$ _____

14. $10 \times 50 =$ _____

15. $20 \times 300 =$ _____

16. $600 \times 70 =$ _____

17. $90 \times 40 =$ _____

Label the number *prime* or *composite*.

18. | 4 | _____

19. | 3 | _____

20. | 11 | _____

Use the Distributive Property to solve.

21. $4 \times 39 = n$

22. $7 \times 26 = n$

23. $5 \times 81 = n$

_____   _____   _____

_____   _____   _____

_____   _____   _____

_____   _____   _____

Use the Distributive Property to solve.

1.

$$5 \times 13 = n$$
$$5 \times (\underline{\hspace{1cm}} + \underline{\hspace{1cm}}) = n$$
$$(5 \times \underline{\hspace{1cm}}) + (5 \times \underline{\hspace{1cm}}) = n$$
$$\underline{\hspace{1cm}} + \underline{\hspace{1cm}} = n$$
$$\underline{\hspace{1cm}} = n$$

2.
$$7 \times 36 = n$$
$$7 \times (\underline{\hspace{1cm}} + \underline{\hspace{1cm}}) = n$$
$$(\underline{\hspace{1cm}} \times \underline{\hspace{1cm}}) + (\underline{\hspace{1cm}} \times \underline{\hspace{1cm}}) = n$$
$$\underline{\hspace{1cm}} + \underline{\hspace{1cm}} = n$$
$$\underline{\hspace{1cm}} = n$$

3.
$$2 \times 19 = n$$
$$2 \times (\underline{\hspace{1cm}} + \underline{\hspace{1cm}}) = n$$
$$(\underline{\hspace{1cm}} \times \underline{\hspace{1cm}}) + (\underline{\hspace{1cm}} \times \underline{\hspace{1cm}}) = n$$
$$\underline{\hspace{1cm}} + \underline{\hspace{1cm}} = n$$
$$\underline{\hspace{1cm}} = n$$

Use the chart to find the answer.

Cereal Sale	
Cereal	Price
Bran Bites	$2.34
Fruity Flakes	$4.45
Organic Corn Clusters	$6.09
Crunchy Wheat Treats	$3.97
Toasted Oats	$2.62

The grocery store has a sale on cereal. Estimate the cost of each purchase.

4. 4 boxes of Toasted Oats _____

5. 7 boxes of Organic Corn Clusters _____

6. 2 boxes of Fruity Flakes _____

7. 6 boxes of Bran Bites _____

8. Complete the chart by listing the cereals and prices from *greatest* to *least* price.

Cereal	Price
	$
	$
	$
	$
	$

Write the number.

9. I have the multiples 12, 18, and 30. I am _____.

10. I have the multiples 25, 60, and 95. I am _____.

11. I am a prime number greater than 5, but less than 10. I am _____.

12. Write multiples of 3. _____, _____, _____, _____, _____

13. Multiples of 10 end in _____.

3-Digit Multipliers

Name _____

Complete the steps to find the product.

1. $60 \times 80 =$ __?__

   Write the basic fact. _____ $\times$ _____ $=$ _____

   Count the 0s in the factors. _____

   Write the product. _____

2. $30 \times 400 =$ __?__

   Write the basic fact. _____ $\times$ _____ $=$ _____

   Count the 0s in the factors. _____

   Write the product. _____

Mark the estimate.

3. $\begin{array}{r} 49 \\ \times 23 \\ \hline ? \end{array}$ ○ 100   ○ 1,000   ○ 10,000

4. $\begin{array}{r} 315 \\ \times\ \ 84 \\ \hline ? \end{array}$ ○ 240   ○ 2,400   ○ 24,000

5. $\begin{array}{r} 297 \\ \times 131 \\ \hline ? \end{array}$ ○ 30,000   ○ 3,000   ○ 300

6. $\begin{array}{r} 408 \\ \times 279 \\ \hline ? \end{array}$ ○ 1,200   ○ 12,000   ○ 120,000

Solve.

7. $\begin{array}{r} 564 \\ \times 823 \\ \hline \end{array}$

8. $\begin{array}{r} \$4.52 \\ \times\ \ \ 236 \\ \hline \end{array}$

9. $\begin{array}{r} 605 \\ \times 182 \\ \hline \end{array}$

10. $\begin{array}{r} 754 \\ \times 321 \\ \hline \end{array}$

11. $\begin{array}{r} 36 \\ \times 18 \\ \hline \end{array}$

12. $\begin{array}{r} 409 \\ \times\ \ 23 \\ \hline \end{array}$

13. $\begin{array}{r} \$12.45 \\ \times\ \ \ \ \ \ 4 \\ \hline \end{array}$

14. $\begin{array}{r} \$1.98 \\ \times\ \ \ \ 12 \\ \hline \end{array}$

Match the property with the example.

_____ 15. $341 \times 0 = 0$

_____ 16. $5 \times 47 = (5 \times 40) + (5 \times 7)$

_____ 17. $29 \times 83 = 83 \times 29$

_____ 18. $156 \times 1 = 156$

_____ 19. $4 \times (5 \times 91) = (4 \times 5) \times 91$

Multiplication Properties

A Associative Property

B Identity Property

C Commutative Property

D Distributive Property

E Zero Property

Label the number *prime* or *composite*.

20. 11 _____

21. 21 _____

22. 23 _____

23. 30 _____

24. 27 _____

25. 51 _____

Solve. Label your answer.

1. Miss Bircher has 19 students. Each student is making fall tree decorations for the classroom. If every student makes 5 leaves, one each of red, yellow, orange, burgundy, and brown paper, how many leaves will the class make?

2. Adriana practices violin for 45 minutes 5 days a week. How many minutes will she have practiced at the end of the 9-week grading period?

3. Troy collects baseball cards. He has 3 boxes with 6 large packages of cards in each box. There are 20 cards in each of the packages. How many cards does he have?

4. Cherie makes carrot cake for her family gatherings. If she uses a 16-ounce bag of carrots for 2 cakes, how many ounces will she use if she makes 4 cakes?

Round the number to the nearest one, tenth, and hundredth.

	one	tenth	hundredth
5. 4.263			
6. 7.956			
7. 3.048			

Match.

_____ 8. 4.137

_____ 9. 4.173

_____ 10. 4.761

_____ 11. 4.278

_____ 12. 4.162

A the greatest number

B $4\frac{162}{1,000}$

C 2 is in the tenths place.

D The value of 3 is 0.03.

E $4 + 0.1 + 0.07 + 0.003$

Write the answer.

13. Write the multiples of 5 up to 20.

_____

14. 16, 40, and 20 are multiples of _____ and _____.

15. I have the multiples of 16, 40, and 72. What number am I?

_____

Complete the table.

16.

Rule: × 4	
Input	Output
20	
50	
70	

17.

Rule: × 9	
Input	Output
3.0	
$10.00	
6,000	

Factor Trees

Complete the factor tree. Write an equation for the number, using only prime factors.

1.
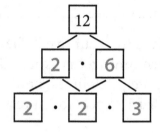

12 = _____

2.
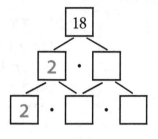

18 = _____

3.
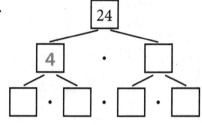

24 = _____

4.

30 = _____

5.

23

23 = _____

6.
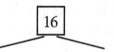

16 = _____

Make a check by the numbers divisible by 2, 5, and 10. Use the divisibility rules.

7.

	2	5	10
30	✓	✓	✓
125			
454			
718			

8.

	2	5	10
56			
27			
98			
380			

Circle the 2 partial products.

9. $4 \times 53 =$ __?__

200

15

12

10. $5 \times 36 =$ __?__

30

300

150

11. $8 \times 29 =$ __?__

720

72

160

12. $9 \times 24 =$ __?__

36

180

18

Solve.

13.
```
   53
 × 34
```

14.
```
  129
 × 75
```

15.
```
  429
 × 58
```

16.
```
   246
 × 199
```

Use the chart to find the answer.

Bruce's Citrus Farm	
Type of Citrus Tree	**Number of Citrus Trees**
orange	585
grapefruit	103
lemon	304
lime	179

1. How many citrus trees does Bruce have?

_____

_____

2. Only 100 of the lime trees survived a heavy winter frost. If each lime tree is expected to produce approximately 220 limes, how many limes can Bruce expect to harvest?

_____

3. Bruce had a bumper crop of grapefruit. He sold bags of 20 grapefruits for $5.00. If his trees produced approximately 100 grapefruits each, approximately how many bags of grapefruit will he be able to sell?

_____

4. Each orange tree produces approximately 250 oranges. Estimate how many oranges the citrus farm will produce.

_____

Calculate the exact amount.

_____

Round to estimate. Solve.

5.
Estimate

$$\begin{array}{r} 415 \\ \times\ 21 \\ \hline \end{array}$$

6.
Estimate

$$\begin{array}{r} 507 \\ \times\ 36 \\ \hline \end{array}$$

Complete the table.

7.
n	$n \cdot 25$
3	
16	
27	
121	

8.
b	$b \cdot 40$
18	
25	
43	
164	

9.
a	$a + 17$
2,153	
1,205	
16,199	
18,457	

Solve.

10. $13 \times 200 =$ _____

11. $61 \times 100 =$ _____

12. $83 \times 20 =$ _____

13. $28 \times 300 =$ _____

14. $133 \times 10 =$ _____

15. $100 \times 100 =$ _____

Multiplication Review

Mark the answer.

1.
- ◯ 4×6
- ◯ $6 + 6 + 6 + 6$
- ◯ 4 sets of 6
- ◯ all of the above

2. $5 + 6 = 6 + 5$
- ◯ Commutative Property
- ◯ Associative Property
- ◯ Distributive Property
- ◯ all of the above

3. $40 \times 600 = \underline{\ ?\ }$
- ◯ 2,400
- ◯ 24,000
- ◯ 240,000
- ◯ all of the above

4. $9 \times 62 = \underline{\ ?\ }$
- ◯ $9 \times (60 + 2)$
- ◯ $(9 \times 60) + (9 \times 2)$
- ◯ $540 + 18$
- ◯ all of the above

5. | 7 | 7 14 21 28
- ◯ factors of 7
- ◯ multiples of 7
- ◯ exponents of 7
- ◯ all of the above

6. $a \cdot b = c$
- ◯ $5 + 4 = 9$
- ◯ $5 - 4 = 1$
- ◯ $5 \times 4 = 20$
- ◯ all of the above

7. 11
- ◯ even number
- ◯ prime number
- ◯ composite number
- ◯ all of the above

Use information from the chart to answer the questions.

School Supplies		
Number per Box	Item	Cost
72	pencils	$4.78
12	colored pencils	$2.99
3	highlighters	$6.49

8. How many boxes of pencils need to be ordered for 150 students to have 1 pencil each?
- ◯ 1 box
- ◯ 3 boxes
- ◯ 5 boxes
- ◯ none of the above

9. What is the cost of 5 boxes of colored pencils?
- ◯ $20.95
- ◯ $30.59
- ◯ $14.95
- ◯ none of the above

10. The school ordered 7 boxes of highlighters. What was the cost?
- ◯ $54.00
- ◯ $40.95
- ◯ $41.53
- ◯ none of the above

11. How many highlighters are in 12 boxes?
- ◯ 34
- ◯ 35
- ◯ 36
- ◯ none of the above

12. What is the cost of 6 boxes of pencils?
- ◯ $28.00
- ◯ $28.68
- ◯ $28.99
- ◯ none of the above

13. How many pencils are in 15 boxes?
- ◯ 720
- ◯ 798
- ◯ 864
- ◯ none of the above

Mark the answer.

14. three sets of seven
- ◯ 3 + 7
- ◯ 7 + 3
- ◯ 3 × 7
- ◯ 7 × 3

15. 3, 6, 9, 12, 15
- ◯ prime numbers
- ◯ composite numbers
- ◯ factors of 15
- ◯ multiples of 3

16. 2, 3, 5, 7
- ◯ prime numbers
- ◯ composite numbers
- ◯ factors of 14
- ◯ multiples of 2

17. 300 × 20 = __?__
- ◯ 6,000
- ◯ 600
- ◯ 60
- ◯ 6

18. 7 × 43 = __?__
- ◯ 7 × 40 × 3
- ◯ 7 × 34
- ◯ 7 × (40 + 3)
- ◯ 7 × 40

19.

- ◯ 2 × 12
- ◯ 2 × 2 × 3
- ◯ 3 × 3 × 3
- ◯ 6 × 3

Mark the answers.

20. Even numbers are divisible by two.
- ◯ 20
- ◯ 41
- ◯ 35
- ◯ 66

21. Numbers ending in 0 or 5 are divisible by 5.
- ◯ 100
- ◯ 445
- ◯ 320
- ◯ 277

22. Numbers ending in 0 are divisible by 10.
- ◯ 3,470
- ◯ 63,450
- ◯ 20,891
- ◯ 8,305

Mark the answer.

23. 5 × 5 × 5
- ◯ 5
- ◯ 5^0
- ◯ 5^1
- ◯ 5^3

24.
$$\begin{array}{r} 43\square \\ \times \quad 5 \\ \hline 2{,}160 \end{array}$$
- ◯ 2
- ◯ 4
- ◯ 6
- ◯ 8

25. 347 × 165 = 165 × __?__
- ◯ 0
- ◯ 1
- ◯ 347
- ◯ 182

Use the data from the chart to find the answer.

City Populations	
Tokyo, Japan	37,468,000
Mexico City, Mexico	21,581,000
New York City, USA	18,819,000
Mumbai, India	19,980,000
Shanghai, China	25,582,000

1. Mark the value of 8 in Tokyo's population.

   A. 80,000,000   C. 8,000,000

   B. 800,000   D. 8,000

2. Mark the greatest place value in Mexico City's population.

   A. Ten Million   C. One Million

   B. Hundred Thousand   D. One Thousand

3. Shanghai's population is less than the population of which city?

   A. Mumbai   C. Shanghai

   B. New York City   D. Tokyo

4. The population of Mumbai can be rounded to which number?

   A. 20,000,000   C. 2,000,000

   B. 200,000   D. 2,000

5. Which city has a population closest to 15,000,000?

   A. Mexico City   C. New York City

   B. Mumbai   D. Shanghai

6. Which 2 cities have about the same population?

   A. Tokyo and Mexico City

   B. New York City and Mumbai

   C. Mexico City and Shanghai

   D. Tokyo and New York City

7. What is the expanded form of the population of Mumbai?

   A. 10,000,00 + 8,000 + 42,000

   B. 1,000,000 + 8,000,000 + 4,000 + 2

   C. 10,000,000 + 9,000,000 + 900,000 + 80,000

   D. 10,000 + 8,000 + 400 + 20

8. What is the word form of the population of New York City?

   A. eighteen million, eight hundred nineteen thousand

   B. sixteen billion, six hundred twenty-six million

   C. sixteen million, six hundred ten thousand, six hundred

   D. sixteen million, six thousand, twenty-six

9. The population of Tokyo can be rounded to which number?

   A. 400,000,000

   B. 300,000,000

   C. 40,000,000

   D. 30,000,000

Mark the answer.

10. The last digit of Ella's phone number is an odd number. What is Ella's phone number?

    A. 342-5700

    B. 342-9842

    C. 342-6421

    D. 342-9536

11. If $a + b = b + a$, then which of these equations is true?

    A. $10 - 6 = 4 + 5$

    B. $9 + 8 = 8 + 9$

    C. $12 - 3 = 9 + 3$

    D. $3 \times 2 = 2 \times 4$

12. If $(7 + 4) + 3 = 7 + (4 + a)$, what is the value of a?

    A. 7

    B. 4

    C. 3

    D. 0

13. What is the rule?

Rule: ?	
Input	Output
6	12
7	13
8	14

    A. subtract 4

    B. multiply by 2

    C. divide by 2

    D. add 6

14. $4.26 + 3.1 = \underline{\ ?\ }$

    A. 7.36

    B. 7.27

    C. 4.57

    D. 1.16

15. Mark the number equal to 472.983.

    A. four hundred seventy-two and nine hundred eighty-three thousandths

    B. $400 + 70 + 2 + 900 + 80 + 3$

    C. $400 + 70 + 2 + 0.003$

    D. $(4 \times 100) + (7 \times 10) + (2 \times 1)$

16. Dylan knows the population of his hometown is about 400,000. What is the exact population?

    A. 472,000

    B. 1,431,850

    C. 4,021,416

    D. 398,430

Use the data from the chart to find the answer.

1-Mile Run	
Runner	Minutes
Anna	8.32
Paul	7.39
Isaac	7.4
Erin	7.9

17. Which runner's time is closest to 8 minutes?

    A. Anna's

    B. Paul's

    C. Isaac's

    D. Erin's

18. Which row shows the times ordered from fastest to slowest?

A. 8.32	7.39	7.4	7.9
B. 7.39	7.4	7.9	8.32
C. 8.32	7.9	7.4	7.39
D. 7.9	7.4	7.39	8.32

Points, Lines & Planes

Name _____

Identify each figure.

line	line segment	plane	point

1.

2.

3.

4.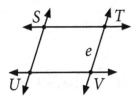

_____   _____   _____   _____

Complete the statement. Use symbols to name the part of plane _e_.

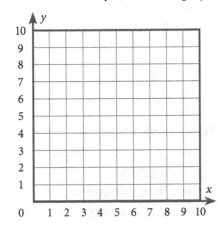

5. A _____ is a flat surface that goes endlessly in all directions.

_____

6. A _____ is a part of a line; it has two endpoints.

_____

7. A _____ is a straight path of points that goes endlessly in both directions.

_____

8. A _____ is a location on a plane.

_____

Plot and label the points on the graph. Answer the question.

9. _A_ (3, 8)   _B_ (5, 7)
   _C_ (7, 6)   _D_ (9, 5)

10. Draw a line through points _A_, _B_, _C_, and _D_.

11. What is the difference between a line and a line segment?

_____

_____

_____

Name a real-life item that resembles the geometric figure.

12. line _____

13. line segment _____

14. point _____

15. plane _____

Draw and label the figure.

16. line _AB_

17. line segment _XY_

18. plane _r_

Circle *prime* or *composite*.

1. | 38 | prime   composite

2. | 57 | prime   composite

3. | 61 | prime   composite

4. | 32 | prime   composite

5. | 29 | prime   composite

6. | 39 | prime   composite

Solve. Label your answer.

> Greenfield Basketball Card Collector's Club has 24 members. Each member has at least one full collection album. Each album holds 20 pages with 12 cards per page.

7. How many cards does each album hold?

8. If each club member has 1 full album, how many pages of cards does the club have in total?

9. What is the smallest number of cards the club owns?

Circle all multiples of the number.

10.	4	8	10	12	14	16	18	20	22	24
11.	5	10	12	15	18	20	25	35	38	40
12.	9	10	18	27	35	40	45	50	54	60

Solve. Label your answer.

13. Rachel's class has 20 students. Each student has 11 textbooks. How many textbooks does her class have?

14. Mrs. Carruthers teaches Spanish to 8 elementary grades. Each grade has 4 classrooms. Each classroom has 23 students. Use the Distributive Property to find out how many students she teaches.

$8 \times ($ _____ $\times$ _____ $) =$

$8 \times ($ _____ $\times 20) + ($ _____ $\times 3) =$

$8 \times ($ _____ $+$ _____ $) =$

$8 \times$ _____ $=$

_____

Solve.

15. $18 \div 6 =$ _____

16. $6\overline{)72}$

17. _____ $\div 6 = 8$

18. _____ $\div 5 = 6$

19. $6 \times 0 \times 6 =$ _____

20. $54 \div 6 =$ _____

21. $\frac{12}{6} =$ _____

22. $6 \times 6 \times 6 =$ _____

23. $(2 \times 3) \cdot 6 =$ _____

Rays & Angles

Name _____

Name the part of the figure.

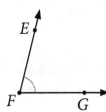

1. Name the vertex. _____

2. Name the two rays. _____

3. Name the angle. _____

Draw an angle. Label the angle ∠ ABC.

4.

Use the figure to name the rays.

5.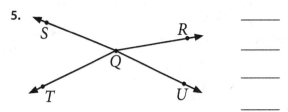

_____

_____

_____

_____

What is the point common to all rays? _____

Label the angle *acute*, *right*, *obtuse*, or *straight*.

6.

_____

7.

_____

8.

_____

9.

_____

Shade the angle. Write the measure of the angle.

10.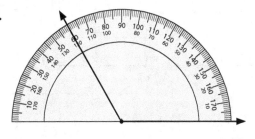

_____

11.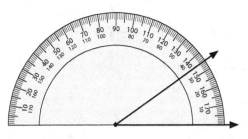

_____

Write the ordered pairs that make the line.

12.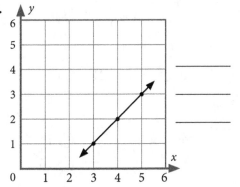

_____

_____

_____

Use plane *f* to find the answers. Write the answers using symbols.

13.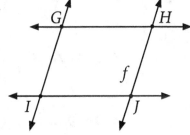

Name 2 lines.

_____

Name 2 points.

_____

Tool Prices	
claw hammer	$9.99
flat-head screwdriver set	$11.99
Phillips-head screwdriver set	$14.99
small drill	$29.93
electric screwdriver	$31.99
measuring tape	$4.99
handsaw	$7.99
circular saw	$34.99
adjustable wrenches	$9.99
gloves	$7.97
socket set	$17.99

Mr. Curtis built a toolbox for each of his four sons. He has $50 to spend to equip each box. He plans to purchase up to five items for each box, and all four boxes must have the same tools.

Use the chart to plan the items for the boxes.

1.

Tool	Price	Price for Four

total price: _____

Write the missing number or letter.

2. $8 \times$ _____ $= 0$

3. _____ $\times 33 = 33$

4. $(5 \times 4) \times 3 =$ _____ $\times (4 \times 3)$

5. $(2 \times$ _____ $) \times 6 = 2 \times (7 \times 6)$

6. $a \cdot (c \cdot d) = (a \cdot$ _____ $) \cdot d$

7. $n \times 0 =$ _____

Write the product.

8. $4 \times 20 =$ _____

9. $5 \times 500 =$ _____

10. $7 \times 2{,}000 =$ _____

Use the Distributive Property to solve.

11.
$$3 \times 29 = \underline{\ ?\ }$$
$$3 \times (\underline{\quad} + \underline{\quad}) =$$
$$(3 \times \underline{\quad}) + (3 \times \underline{\quad}) =$$
$$\underline{\quad} + \underline{\quad} = \underline{\quad}$$

12.
$$7 \times 45 = \underline{\ ?\ }$$
$$7 \times (\underline{\quad} + \underline{\quad}) =$$
$$(\underline{\quad} \times \underline{\quad}) + (\underline{\quad} \times \underline{\quad}) =$$
$$\underline{\quad} + \underline{\quad} = \underline{\quad}$$

13.
$$5 \times 16 = \underline{\ ?\ }$$
$$\underline{\quad} \times (\underline{\quad} + \underline{\quad}) =$$
$$(\underline{\quad} \times \underline{\quad}) + (\underline{\quad} \times \underline{\quad}) =$$
$$\underline{\quad} + \underline{\quad} = \underline{\quad}$$

14.
$$8 \times 62 = \underline{\ ?\ }$$
$$8 \times (\underline{\quad} + \underline{\quad}) =$$
$$(\underline{\quad} \times \underline{\quad}) + (\underline{\quad} \times \underline{\quad}) =$$
$$\underline{\quad} + \underline{\quad} = \underline{\quad}$$

Complete the equation.

15. $7 \times 7 = \underline{\ ?\ }$ ○ 59 ○ 49

16. $\underline{\ ?\ } \times 7 = 42$ ○ 6 ○ 8

17. $\underline{\ ?\ } \times 7 = 21$ ○ 3 ○ 4

18. $7 \times 8 = \underline{\ ?\ }$ ○ 54 ○ 56

19. $7 \times 9 = \underline{\ ?\ }$ ○ 63 ○ 64

20. $7 \times 2 = \underline{\ ?\ }$ ○ 14 ○ 16

Measuring Angles

Identify the lines as *intersecting*, *parallel*, or *perpendicular*.

1.

_____

2.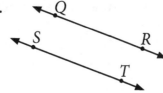

_____

3.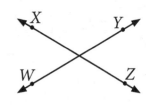

_____

Use the figure to find the answer.

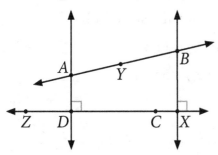

4. Name two lines that are parallel. _____

5. Name two lines that are perpendicular.

_____

6. Name a point on line *AB*. _____

7. Name a right angle. _____

Use the figure to name an angle and classify it as *acute*, *obtuse*, *right*, or *straight*.

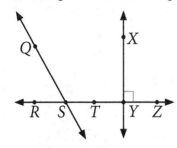

8. It measures greater than 90°. _____ is _____.

9. It measures less than 90°. _____ is _____.

10. It measures 90°. _____ is _____.

11. It measures 180°. _____ is _____.

Use a symbol and three points to name the angle. Classify the angle as *acute*, *obtuse*, *right*, or *straight*.
Write the measure of the angle.

12.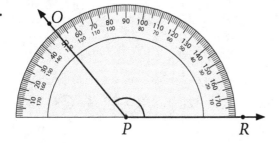

name: _____

classification: _____

measure: _____

13.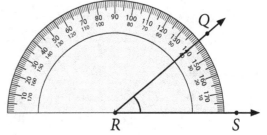

name: _____

classification: _____

measure: _____

Use plane _h_ to find the answer.

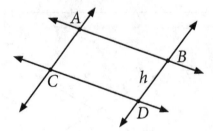

1. Name 2 points. _____

2. Name 2 lines. _____

3. Name 2 line segments. _____

Plot and label the points on the graph. Use the graph to find the answer.

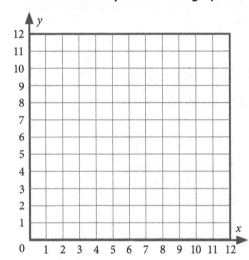

4. _A_ (5, 2)   _B_ (3, 1)   _C_ (1, 7)
   _D_ (4, 8)   _E_ (10, 8)

5. Do points _A, B, C, D,_ and _E_ form a straight line? _____

6. Do any of the points share an _x_-coordinate? _____

7. Do any of the points share a _y_-coordinate? _____

Solve.

8. $417 + 33 =$ _____

9. $6.73 + 2.26 =$ _____

10. $\$10.99 + \$3.67 =$ _____

11. $\begin{array}{r} \$56.11 \\ +\ \$1.99 \\ \hline \end{array}$

12. $\begin{array}{r} 6{,}522 \\ +3{,}468 \\ \hline \end{array}$

13. $\begin{array}{r} 5.631 \\ +4.784 \\ \hline \end{array}$

14. $\begin{array}{r} 6{,}561{,}214 \\ +\ 1{,}384{,}697 \\ \hline \end{array}$

15. $\frac{32}{4} =$ _____

16. $8\overline{)80}$

17. $8 \times$ _____ $= 16$

18. _____ $\times\ 8 = 72$

Mark the answer.

19. $8 \times 7 =$?   ○ 52   ○ 54   ○ 56

20. $8 \times 8 =$?   ○ 62   ○ 63   ○ 64

21. $6 \times 8 =$?   ○ 42   ○ 46   ○ 48

22. $3 \times 8 =$?   ○ 22   ○ 24   ○ 26

23. $5 \times 8 =$?   ○ 35   ○ 40   ○ 45

24. $8 \times 0 =$?   ○ 0   ○ 1   ○ 8

Measuring & Drawing Angles

Name _____

Mark the best measurement. Classify the angle.

1.

- ○ 60°   ○ acute
- ○ 180°   ○ obtuse

2.

- ○ 80°   ○ acute
- ○ 100°   ○ obtuse

3.
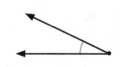

- ○ 25°   ○ acute
- ○ 155°   ○ obtuse

Draw and label the figure. Use a ruler and protractor.

4. ∠ABC

5. $\overleftrightarrow{ST}$ is parallel to $\overleftrightarrow{UV}$.

6. $\overline{MN}$ is perpendicular to $\overline{OP}$.

Complete the equation to find the measure of the unknown angle.

7.
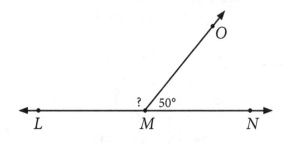

∠OMN = 50°   ∠OML = _?_

$$\underline{\quad 50° \quad} + n = 180°$$
$$n = 180° - \underline{\quad\quad}$$
$$n = \underline{\quad\quad}$$
$$\angle OML = \underline{\quad\quad}$$

8.
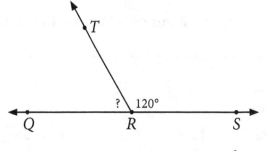

∠TRS = 120°   ∠TRQ = _?_

$$\underline{\quad 120° \quad} + n = 180°$$
$$n = 180° - \underline{\quad\quad}$$
$$n = \underline{\quad\quad}$$
$$\angle TRQ = \underline{\quad\quad}$$

Plot and label the points on the graph.
Connect the points to form a line.

1. A (8, 1)   B (6, 3)
   C (4, 5)   D (2, 7)

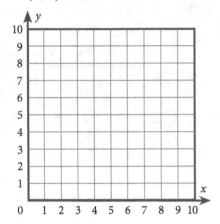

Estimate by rounding. Circle three items.

2. Mr. Matthews needs new lawn and yard equipment. He can spend $180. What 3 items could he purchase?

   gas-powered edger $149.95

   garden hoe $22.75

   electric edger $85.00

   electric leaf blower $63.29

   gas leaf blower $114.99

Use the chart to answer the question.

Prices of School Supplies	
backpack	$25.99
3-ring binder	$14.79
calculator	$5.79
pencil box with supplies	$7.86

3. How much does it cost to purchase a calculator and a pencil box with supplies? _____

4. How much does it cost to purchase a backpack and a calculator? _____

5. How much do all of the supplies cost? _____

6. Would $35 be enough to purchase all of these supplies? _____

   Why or why not? _____

7. Which 2 items could be purchased with $20?

   _____

Solve.

8. 246 + 13 = _____

9. 3,456 − 165 = _____

10. $364.13 + $11.01 = _____

Write the missing number.

11. ⬚) 81 quotient 9

12. 9) 54 quotient ⬚

13. 9) ⬚ quotient 7

14. 3) ⬚ quotient 9

15. ⬚) 18 quotient 2

16. $\frac{54}{9}$ = ⬚

17. $\frac{72}{9}$ = ⬚

18. $\frac{36}{⬚}$ = 4

19. $\frac{45}{5}$ = ⬚

20. $\frac{⬚}{9}$ = 3

Triangles

Name _____

Complete the equation to find the measure of the unknown angle.

1.

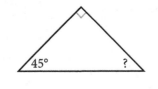

$180° - ($ _____ $+$ _____ $) =$ __?__

$180° -$ _____ $=$ _____

2.

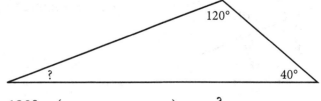

$180° - ($ _____ $+$ _____ $) =$ __?__

$180° -$ _____ $=$ _____

3.

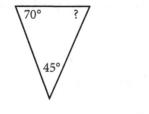

$180° - ($ _____ $+$ _____ $) =$ __?__

$180° -$ _____ $=$ _____

4.

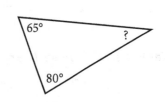

$180° - ($ _____ $+$ _____ $) =$ __?__

$180° -$ _____ $=$ _____

Classify the triangle as *acute*, *right*, or *obtuse*.

5.

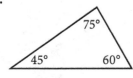

_____

6.

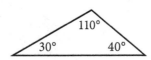

_____

7.

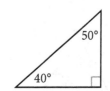

_____

8.

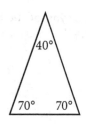

_____

Find the measure of the third angle.

9. Two angles of a triangle measure 50° each.
   What is the measure of the third angle? _____

10. A triangle has angles that measure 25° and 55°.
    What is the measure of the third angle? _____

11. Two angles of a triangle measure 90° and 60°.
    What is the measure of the third angle? _____

Use plane *m* to find the answer.

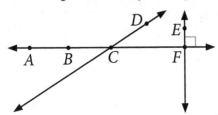

1. $\overleftrightarrow{QR}$ is parallel to _____.

2. $\overleftrightarrow{QR}$ is perpendicular to _____.

3. $\overleftrightarrow{QR}$ intersects with _____.

Use the figure to classify the angle as *acute, obtuse, right,* or *straight.*

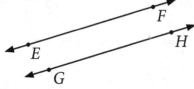

4. $\angle BCD$ is _____. It measures greater than 90°.

5. $\angle DCF$ is _____. It measures less than 90°.

6. $\angle EFC$ is _____. It measures 90°.

7. $\angle FCB$ is _____. It measures 180°.

Identify the lines as *intersecting, parallel,* or *perpendicular.*

8.
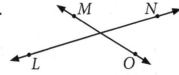

_____

9.

_____

10.
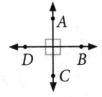

_____

Use a symbol and three points to name the angle. Classify the angle as *acute, obtuse, right,* or *straight.*
Write the measure of the angle.

11.
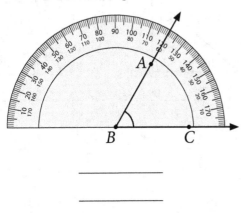

_____

_____

_____

12.
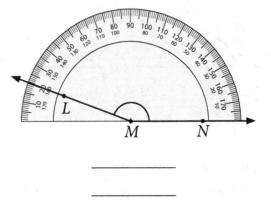

_____

_____

_____

Solve.

13. $6 \times$ _____ $= 54$

14. $49 \div$ _____ $= 7$

15. $5 \times 6 =$ _____

16. $9 \times 7 =$ _____

17. $7 \times$ _____ $= 42$

18. $36 \div$ _____ $= 6$

19. $24 \div 6 =$ _____

20. $54 \div 9 =$ _____

21. _____ $\times 2 = 14$

22. _____ $\times 5 = 35$

23. $48 \div 8 =$ _____

24. $72 \div 8 =$ _____

Circles

Follow the directions to label the circle.

1. Label the circle R.

2. Draw radius $\overline{RS}$.

3. Draw diameter $\overline{MN}$.

4. Draw chord $\overline{PQ}$.

5. Name an angle formed by the diameter and radius.

   _____

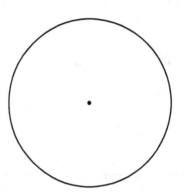

Write the measurement and label.

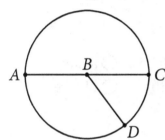

$\overline{BD}$ = 7 in.

6. $\overline{BC}$ = _____

7. $\overline{AC}$ = _____

8. Name the circle. _____

Explain and solve.

9. Terra knows the diameter of a can is 6 inches. How can she find the radius?

   _____

Plot and label the points on the graph. Connect the points.

10. A (2, 5)   11. B (4, 4)

12. C (6, 3)   13. D (8, 2)

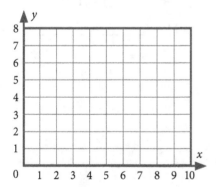

Bonus:

What would be a point on the extended line?

_____

Complete the equation to find the measure of the unknown angle.

14.

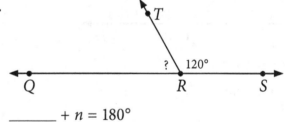

_____ + n = 180°

n = 180° − _____

$\angle TRQ$ = _____

15.

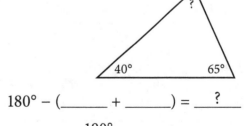

180° − (_____ + _____) = ___?___

180° − _____ = _____

Classify the angle as *acute, obtuse, right,* **or** *straight.* **Use a protractor. Write the measure of the angle.**

1.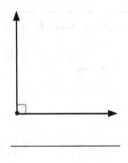

_____

_____

2.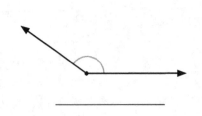

_____

_____

3.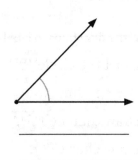

_____

_____

Mark the best measurement. Classify the angle.

4.

○ 95°   ○ acute
○ 35°   ○ obtuse

5.

○ 75°   ○ acute
○ 25°   ○ obtuse

6.

○ 90°   ○ acute
○ 130°   ○ obtuse

Complete the equation to find the measure of the unknown angle.

7.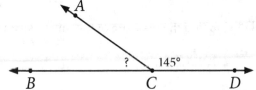

$145° + n = 180°$

$n = 180° -$ _____

$\angle ACB =$ _____

8.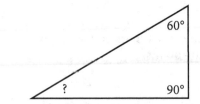

$180° - ($ _____ $+$ _____ $) = 180°$

$180° -$ _____ $=$ _____

Complete the multiplication table.

9.

×	1	2	3	4	5	6	7	8	9	10	11	12
8	8	16										

10.

×	1	2	3	4	5	6	7	8	9	10	11	12
9	9	18										

Graphing Figures

Name _____

Plot and label the points on the graph. Draw the figure and complete the statement.

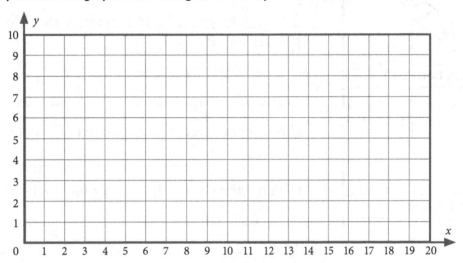

Figure 1

1. Q (4, 3)
   R (10, 7)
   Draw $\overrightarrow{QR}$.

2. S (5, 5)
   T (10, 5)
   Draw $\overleftrightarrow{ST}$.

3. $\overrightarrow{QR}$ and $\overleftrightarrow{ST}$ are _____ lines.

4. Z (7, 5)

5. Name an acute angle, _____,
   an obtuse angle, _____, and
   a straight angle _____.

6. If $\angle SZQ$ is 35°, what is the measurement
   of $\angle QZT$? _____

Figure 2

7. U (15, 2)
   V (19, 2)
   Draw $\overrightarrow{UV}$.

8. W (17, 1)
   X (17, 4)
   Draw $\overleftrightarrow{WX}$.

9. $\overrightarrow{UV}$ and $\overleftrightarrow{WX}$ are _____ lines.

10. Draw a box at each 90° angle.

Follow the directions.

11. Label the circle L.

12. Draw diameter $\overline{MN}$.

13. Draw radius $\overline{LP}$.

14. Draw chord $\overline{JK}$.

15. Name an angle formed by the diameter
    and radius. _____

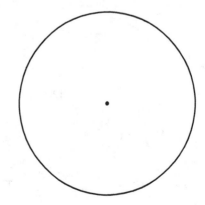

Use the map to find the answer.

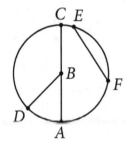

1. Name the street that is parallel to Lehman Lane.

   _____

2. Name the streets that intersect Satterfield Avenue and Brad Street.

   _____

   _____

3. Name the street that is parallel to Satterfield Avenue.

   _____

4. Name the streets that form right angles.

   _____

   _____

   _____

   _____

Use symbols to name the parts of circle B.

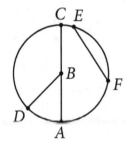

5. radii _____ _____ _____

6. diameter _____

7. chord _____

Complete the equation to find the measure of the unknown angle.

8.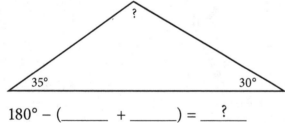

   $180° - ($ _____ $+$ _____ $) =$ __?__

   $180° - $ _____ $=$ _____

9.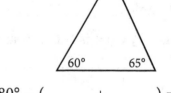

   $180° - ($ _____ $+$ _____ $) =$ __?__

   $180° - $ _____ $=$ _____

Write the missing number.

10. $9 \overline{) 81}$

11. $9 \overline{) 54}$

12. $9 \overline{) }\ \ 7$

13. $3 \overline{) }\ \ 9$

14. $ \overline{) 18}\ \ 2$

15. $\frac{18}{2} =$ _____

16. $3 \times 6 =$ _____

17. $5 \times$ _____ $= 40$

18. $4 \times 9 =$ _____

© 2021 BJU Press. Reproduction prohibited.

Geometry Review

Name _____

Mark the answer.

1.
   - ○ acute
   - ○ obtuse
   - ○ right

2.
   - ○ obtuse
   - ○ right
   - ○ straight

3.
   - ○ obtuse
   - ○ right
   - ○ straight

4.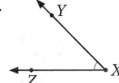
   - ○ acute
   - ○ obtuse
   - ○ right

5.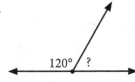
   - ○ ∠XYZ
   - ○ ∠ZYX
   - ○ ∠ZXY

6.
   120° ?
   - ○ 180°
   - ○ 60°
   - ○ 90°

7.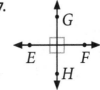
   - ○ plane
   - ○ parallel lines
   - ○ perpendicular lines

8.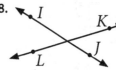
   - ○ intersecting lines
   - ○ parallel lines
   - ○ perpendicular lines

9.
   - ○ parallel lines
   - ○ perpendicular lines
   - ○ plane

10.
   - ○ parallel lines
   - ○ perpendicular lines
   - ○ plane

11.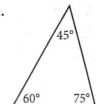
    25° 130° 25°
   - ○ acute triangle
   - ○ obtuse triangle
   - ○ right triangle

12.
    45° 60° 75°
   - ○ acute triangle
   - ○ obtuse triangle
   - ○ right triangle

Use the circle for 13–17. Mark all correct answers.

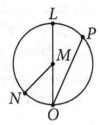

Mark the answer.

18.

$$180° - (60° + 60°) = \underline{\ ?\ }$$

○ 60°
○ 70°
○ 80°
○ 90°

13. radius

○ $\overline{ML}$
○ $\overline{MN}$
○ $\overline{MO}$
○ $\overline{OP}$

19.

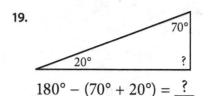

$$180° - (70° + 20°) = \underline{\ ?\ }$$

○ 60°
○ 70°
○ 80°
○ 90°

14. diameter

○ $\overline{MN}$
○ $\overline{OL}$
○ $\overline{OP}$
○ $\overline{LO}$

20.

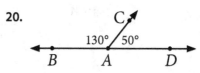

$$\angle BAC + \angle DAC = \underline{\ ?\ }$$

○ 90°
○ 100°
○ 180°
○ 360°

15. chord

○ $\overline{NM}$
○ $\overline{OP}$
○ $\overline{LO}$
○ $\overline{OM}$

21.

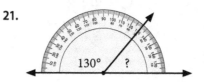
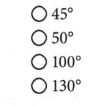

○ 45°
○ 50°
○ 100°
○ 130°

Use the graph for 22–23. Mark the correct answer.

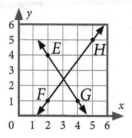

16. circle

○ circle O
○ circle N
○ circle M
○ circle L

17. central angle

○ $\angle NMO$
○ $\angle NML$
○ $\angle MOP$
○ $\angle POL$

22. Name the point at (2, 4).

○ E
○ F
○ G
○ H

23. Name the coordinates for H.

○ (2, 1)
○ (5, 3)
○ (5, 5)
○ (3, 3)

Mark the related fact.

1. | $16 - 7 = 9$

   A. $8 + 8 = 16$
   B. $9 + 7 = 16$
   C. $16 - 8 = 8$
   D. $9 - 7 = 2$

2. | $14 - 6 = 8$

   A. $6 + 8 = 14$
   B. $8 + 6 = 14$
   C. $14 - 8 = 6$
   D. all of the above

3. | $12 - 3 = 9$

   A. $9 - 3 = 6$
   B. $12 - 6 = 6$
   C. $3 + 9 = 12$
   D. none of the above

4. | $17 - 8 = 9$

   A. $4 + 4 = 8$
   B. $8 + 9 = 17$
   C. $10 + 7 = 17$
   D. $9 - 1 = 8$

5. | $11 - 5 = 6$

   A. $11 - 6 = 5$
   B. $5 + 5 = 10$
   C. $7 + 4 = 11$
   D. $3 + 8 = 11$

Mark the related fact.

6. | $12 = \underline{\ ?\ }$

   A. $2 \times 2 \times 3$
   B. $2 \times 3 \times 3$
   C. $2 \times 2 \times 5$
   D. $2 \times 3 \times 5$

7. | $36 = \underline{\ ?\ }$

   A. $2 \times 2 \times 5$
   B. $2 \times 3 \times 3$
   C. $2 \times 2 \times 3 \times 3$
   D. $3 \times 3 \times 5$

8. | $75 = \underline{\ ?\ }$

   A. $2 \times 2 \times 5$
   B. $2 \times 2 \times 2 \times 3$
   C. $3 \times 3 \times 5$
   D. $3 \times 5 \times 5$

9. | $25 = \underline{\ ?\ }$

   A. $2 \times 5 \times 5$
   B. 5×5
   C. $2 \times 2 \times 2 \times 3$
   D. $2 \times 3 \times 5$

10. | $17 = \underline{\ ?\ }$

    A. 1×17
    B. $2 \times 3 \times 3$
    C. 3×5
    D. $2 \times 2 \times 5$

11. What is true about the set of numbers?

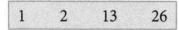

| 1 | 2 | 13 | 26 |

A. Only 2 and 13 are prime numbers.
B. All are factors of 26.
C. The number 26 is a composite number.
D. All of the above are true.

12. What factor of 36 is missing?

36: 1, 2, 3, 4, 6, __?__, 12, 18, 36

A. 7   C. 9
B. 8   D. 10

13. Which is *not* a name for 24?

A. 6 × 4   C. 3 × 8
B. 12 + 12   D. 10 + 12

14. Which is *not* a name for 1,596?

A. 1,000 + 500 + 90 + 6
B. one thousand, five hundred ninety-six
C. 1.596
D. 3 × 532

15.

What does the picture show?

A. 3 + 3 + 3   C. 3 × 8
B. 3 in each set of 7   D. 3 sets of 7

16.

45 ft

45 ft [ ] 45 ft

45 ft

What is the sum of the sides?

A. 140 ft   C. 180 ft
B. 165 ft   D. 200 ft

17.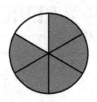

What part of the circle is shaded?

A. $\frac{5}{6}$ C. $\frac{1}{7}$

B. $\frac{1}{6}$ D. $\frac{5}{7}$

18.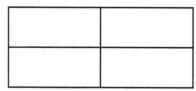

How many parts should be colored to show $\frac{3}{4}$ of the rectangle?

A. 4   C. 2
B. 3   D. 1

19.

What part of the set is airplanes?

A. $\frac{1}{2}$ C. $\frac{2}{4}$

B. $\frac{3}{5}$ D. none of these

20.

The dotted line shows the first fold of a paper airplane. What will the paper look like when it is folded?

A. [rectangle]   C. [rectangle]

B. [triangle]   D. none of these

Division

Name _____

Write the division equation for the phrase. Solve. Draw a picture to illustrate.

1. 35 stamps divided onto 5 pages

2. 18 juice boxes divided into 6 packs

3. 1 dozen muffins divided among 6 family members

_____ _____ _____

Solve and label your answer.

4. Suzanna has 3 packs of gum. She has a total of 24 pieces of gum. How many pieces are in each pack?

5. Ethan has 32 model cars in his collection. He divides his cars evenly among 4 shelves. How many cars are displayed on each shelf?

Write the quotient.

6. $8\overline{)64}$

7. $9\overline{)45}$

8. $8\overline{)48}$

9. $6\overline{)54}$

10. $6\overline{)36}$

11. $\frac{28}{4} =$ _____

12. $\frac{81}{9} =$ _____

13. $\frac{42}{6} =$ _____

14. $\frac{56}{7} =$ _____

15. $\frac{72}{8} =$ _____

Write a division equation to find the missing factor. Solve for _n_.

16. $5 \times n = 45$

$n = 45 \div 5$

$n =$ _____

17. $7 \times n = 63$

_____

_____

18. $2 \times n = 12$

_____

_____

Complete the chart by drawing shells to find the answer.

19. Cherith has 35 shells that she collected from her trip to the beach. She went to the beach 5 days and collected the same number each day. How many did she collect each day?

Monday	Tuesday	Wednesday	Thursday	Friday

Write the division problem 3 ways.

1.
> The dividend is 35.
> The divisor is 5.
> The quotient is 7.

_____

2.
> The dividend is 48.
> The divisor is 8.
> The quotient is 6.

_____

Answer the riddle.

3. I am a whole number that can be divided into 4 equal pieces. I am greater than 10 and less than 16. _____

4. I am a number divisible by 5. My value is less than 100, and the sum of my digits is 11. _____

5. I am a number that is a multiple of 10. I am divisible by 2, 5, and 6. I am less than 60. _____

Use the figure to answer the questions.

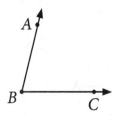

6. Name the vertex. _____

7. Name the two rays. _____

8. Name the angle. _____

Label the angle as *acute, right, obtuse,* **or** *straight.*

9.

10.

11.

12.

_____   _____   _____   _____

Write the missing number.

13.
$9\overline{)81}$

14.
$\overline{)56}\;8$

15.
$7\overline{)49}$

16.
$\overline{)45}\;9$

17.
$8\overline{)}\;9$

Solve.

18. $\frac{40}{5} =$ _____

19. $\frac{21}{7} =$ _____

20. $\frac{16}{2} =$ _____

21. $\frac{36}{4} =$ _____

22. $\frac{12}{3} =$ _____

1-Digit Quotients

Name _____

Solve.

1.

2.

3.

4.

Find the quotient. Use multiplication to check.

5. 9)7 4

Check

6. 7)2 3

Check

Solve and label your answer.

7. Mrs. Moore's class has 19 students. Her husband built a cubby unit 5 rows high with 5 cubby spaces on each row. How many cubbies does the whole unit have?

_____

If each student is assigned one cubby, how many cubbies will be empty?

_____

Write a division equation and a multiplication equation.
Complete the statement to explain how the two equations are related.

8.

35
5	5	5	5	5	5	5

_____

_____

There are _____ sets of _____ in _____.

9.

64
8 + 8 + 8 + 8 + 8 + 8 + 8 + 8

_____

_____

There are _____ sets of _____ in _____.

Solve for n.

10.
$$\begin{array}{r} 5 \\ \times\, n \\ \hline 15 \end{array}$$
$n =$ _____

11.
$$\begin{array}{r} 3 \\ \times\, n \\ \hline 27 \end{array}$$
$n =$ _____

12. $4 \times n = 28$

$n =$ _____

13. $\frac{54}{n} = 9$

$n =$ _____

Complete the equation to find the measure of the unknown angle.

1.
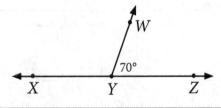

$\angle WYZ = 70°$ $\angle XYW = \underline{?}$

_____ + n = 180°

n = 180° – _____

n = _____

$\angle XYW$ = _____

2.
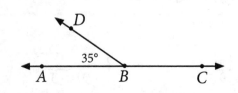

$\angle ABD = 35°$ $\angle DBC = \underline{?}$

_____ + n = 180°

n = _____ – _____

n = _____

$\angle DBC$ = _____

Estimate the product.

3.
$$\begin{array}{r} 63 \\ \times\ 4 \\ \hline \end{array}$$

4.
$$\begin{array}{r} 316 \\ \times\ 2 \\ \hline \end{array}$$

5.
$$\begin{array}{r} 437 \\ \times\ 5 \\ \hline \end{array}$$

6.
$$\begin{array}{r} 2,951 \\ \times\ 6 \\ \hline \end{array}$$

7.
$$\begin{array}{r} 8,777 \\ \times\ 5 \\ \hline \end{array}$$

Solve.

8.
$$\begin{array}{r} \$4.26 \\ \times\ 3 \\ \hline \end{array}$$

9.
$$\begin{array}{r} \$7.19 \\ \times\ 5 \\ \hline \end{array}$$

10.
$$\begin{array}{r} 3,400 \\ \times\ 6 \\ \hline \end{array}$$

11.
$$\begin{array}{r} \$9.99 \\ \times\ 8 \\ \hline \end{array}$$

12.
$$\begin{array}{r} \$1.36 \\ \times\ 5 \\ \hline \end{array}$$

Solve and label your answer.

13. Mrs. Smith spent $14.67 for each of her 3 children's lunch boxes. How much did she spend for all 3?

14. Lucy has 77 blank recipe cards. She divided her recipe box into 9 sections. If 5 sections have 1 more card than all the other sections, how many cards are in each section?

Solve.

15.	$300 \times 10 = \underline{?}$	300	3,000	30,000
16.	$50 \times 20 = \underline{?}$	100	1,000	10,000
17.	$600 \times 800 = \underline{?}$	4,800	48,000	480,000
18.	$10 \times 7,000 = \underline{?}$	7,000	70,000	700,000
19.	$50 \times 700 = \underline{?}$	350	3,500	35,000
20.	$1,000 \times 20 = \underline{?}$	2,000	20,000	200,000

21.	$54 \div 9 = \underline{?}$	6	7	8
22.	$56 \div 7 = \underline{?}$	7	8	9
23.	$49 \div 7 = \underline{?}$	6	7	8
24.	$36 \div 9 = \underline{?}$	4	5	6
25.	$63 \div 7 = \underline{?}$	7	8	9
26.	$72 \div 9 = \underline{?}$	7	8	9

1- & 2-Digit Quotients

Name _____

Write the equation. Solve. Write the answer.

1. 78 band members in 6 rows

_____

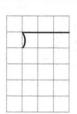

_____ band members in each row

2. 130 fifth graders in 5 classes

_____

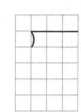

_____ fifth graders in each class

3. 208 pages in 8 folders

_____

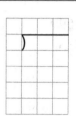

_____ pages in each folder

Solve and label your answer.

4. Esther has 168 marshmallows divided equally among 3 bags. How many are in each bag?

5. Robin has 17 carrots to feed her pet rabbits. If she plans to feed the rabbits 3 carrots per day, will she have enough to last 1 week?

6. Alex helped his father rake the leaves. They bagged up 15 bags and stacked them in 4 neat rows. Could all of the rows have the same number of bags? Explain your answer.

7. William has $6.00 to spend on treats for his 3 friends. If he spends the same amount on each friend, what is the maximum amount of money he can spend on each person?

Solve.

8.
$8\overline{)162}$

9.
$5\overline{)344}$

10.
$6\overline{)610}$

11.
$9\overline{)127}$

Circle the problem with the correct answer.

12. $\begin{array}{r}13\,r2\\3\overline{)41}\end{array}$ or $\begin{array}{r}14\\3\overline{)41}\end{array}$

13. $\begin{array}{r}40\,r2\\6\overline{)246}\end{array}$ or $\begin{array}{r}41\\6\overline{)246}\end{array}$

Complete the table.

1.

n	n + 6
52	
98	
241	
699	
900	

2.

n	n · 2
15	
36	
73	
115	
610	

3.

n	n ÷ 7
49	
147	
210	
497	
707	

Write the value of *n*.

4. $(4 + n) + 9 = 4 + (5 + 9)$

   $n = \underline{\hspace{1cm}}$

5. $619 + n = 619$

   $n = \underline{\hspace{1cm}}$

6. $n + 34 = 17 + 34$

   $n = \underline{\hspace{1cm}}$

7. $5 + n = 11$

   $n = \underline{\hspace{1cm}}$

8. $n - 56 = 43$

   $n = \underline{\hspace{1cm}}$

9. $\frac{39}{n} = 3$

   $n = \underline{\hspace{1cm}}$

Round to the nearest whole number to estimate. Solve.

10. **Estimate**
$$\begin{array}{r} 6.411 \\ + 3.567 \\ \hline \end{array}$$

11. **Estimate**
$$\begin{array}{r} 5.963 \\ + 4.281 \\ \hline \end{array}$$

12. **Estimate**
$$\begin{array}{r} 9.146 \\ + 1.003 \\ \hline \end{array}$$

Round to the nearest thousand. Estimate and label the answer.

13. Mrs. Dover's 6th-grade class collected 2,300 cans for recycling in February; 5,800 in March; 4,917 in April; and 5,832 in May. About how many cans did they collect for recycling in those 4 months?

_____

Solve.

14. $\begin{array}{r} 137 \\ \times\ 15 \\ \hline \end{array}$

15. $\begin{array}{r} 436 \\ \times\ 50 \\ \hline \end{array}$

16. $\begin{array}{r} 518 \\ \times\ 11 \\ \hline \end{array}$

17. $\begin{array}{r} 560 \\ \times\ 28 \\ \hline \end{array}$

18. $712 - 99 =$ _____

19. $437 - 29 =$ _____

20. $100{,}000 - 10{,}000 =$ _____

2- & 3-Digit Quotients

Name _____

Complete the table. Use the table to answer the questions.

> Max delivers products from his farm each week. He wants to find the average miles he travels.

Trips in March			
Destination	Total Miles Traveled	Number of Deliveries	Average Miles per Delivery
farmers' market	540	6	
grocery store	256	8	
wholesale warehouse	135	9	
soup kitchen	905	5	

1. How many deliveries did Max make in March? _____

2. How many miles is one trip to the farmers' market and back to the farm?

   _____

3. How many total miles did Max travel in March? _____

4. If Max makes the same trips each month, how many miles will he travel in a year?

   _____

5. Which delivery place is farthest from the farm? _____

6. Which delivery place is closest to the farm? _____

Use Max's map to answer the questions.

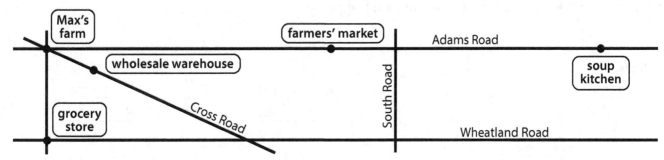

7. Name the road parallel to Adams Road. _____

8. Name the road perpendicular to Adams Road. _____

9. Name a road that intersects Wheatland Road. _____

Solve. Use multiplication to check.

10. 7)2 9 5   Check

11. 4)6 5 5   Check

Write the value of the underlined digit.

1. 4<u>4</u>3,216,503 _____
2. 671,<u>9</u>04,543 _____
3. 950,47<u>8</u>,009 _____
4. 99,<u>9</u>99,999 _____
5. <u>1</u>27,549 _____
6. 5<u>9</u>6,208 _____

Write >, <, or = to compare.

7. 347 ◯ 344
8. 23,765 ◯ 37,143
9. 54,632 ◯ 50,000 + 4,000 + 600 + 30 + 2
10. 1,791.450 ◯ 1,791,540
11. 2,004 ◯ twenty-four thousand, four
12. 3,607 ◯ three thousand, six hundred seven

Write the decimal.

13. $15\frac{36}{100}$ = _____
14. $3\frac{6}{100}$ = _____
15. eighty-nine thousandths = _____
16. three and three tenths = _____

Circle the digit in the given place.

17.	Hundredths place	5 6 . 1 3 8
18.	Ones place	1 1 2 . 4 0 9
19.	Tenths place	9 7 . 6 2 1
20.	Tens place	8 3 . 5 0 5

Use plane *m* to find the answer.

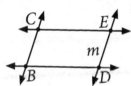

21. Name two lines that intersect with $\overleftrightarrow{BC}$. _____
22. Name an acute angle. _____
23. Name a point on the plane. _____

Multiply.

24. 213 × 61
25. 451 × 37
26. 1,609 × 42
27. $12.75 × 25

Divide.

28. 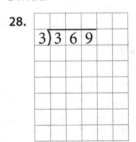 3)3 6 9

29. 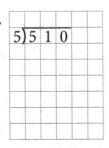 5)5 1 0

30. 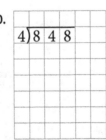 4)8 4 8

31. 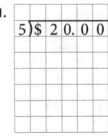 5)$ 2 0 . 0 0

32. 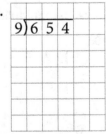 9)6 5 4

Zero in the Quotient

Name _____

Solve.

1. 4)411

2. 8)801

3. 3)59

4. 2)175

Complete the table.

5.
Rule: ÷ 5	
Input	Output
25	
55	
500	
550	
1,000	

6.
Rule: ÷ 3	
Input	Output
36	
39	
63	
93	
99	

7.
Rule: ÷ 2	
Input	Output
20	
46	
64	
88	
102	

Write a division equation for the phrase. Solve. Write the answer.

8. 84 juice boxes in packs of 6

_____

_____ 6-packs of juice boxes

9. 165 students in 5 grades

_____

average of _____ students per grade

Solve and label your answer.

10. Mrs. Lewis purchased fruit snacks for her son Trey's class. She bought enough for each of the 23 students to have 2 packs. If the snacks are sold in 8-pack boxes, how many boxes did she buy?

Write a division equation to find the missing factor. Solve for *n*.

11. $n \times 7 = 49$

$n = 49 \div 7$

$n = $ _____

12. $3 \times n = 21$

_____

_____

13. $n \times 15 = 105$

_____

_____

14. $8 \times n = 104$

_____

_____

Solve.

1. 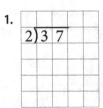 $2\overline{)37}$

2. 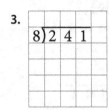 $3\overline{)203}$

3. $8\overline{)241}$

4. $7\overline{)316}$

Use the table to answer.

Amanda is helping her mother prepare lunches for a picnic. Choose your lunch from the items below. Keep the cost at $10.00 or less.

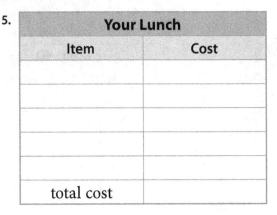

Lunch Items	Price
turkey sub	$6.75
ham sub	$6.50
chips	$1.10
apple	$1.50
water	$1.50
juice	$1.80
cookie	$0.65

5.

Your Lunch	
Item	Cost
total cost	

Write the decimal that goes in each blank.

6.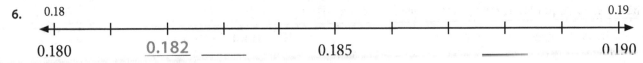

0.18   0.19

0.180   0.182 _____   0.185   _____   0.190

7.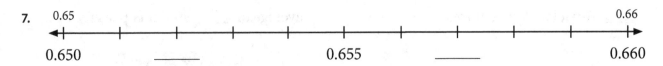

0.65   0.66

0.650   _____   0.655   _____   0.660

Write the numbers from *least* to *greatest*.

8. ⁻5   2   4   ⁻1

_____ _____ _____ _____

9. ⁻13   ⁻18   15   ⁻31

_____ _____ _____ _____

Complete the table.

10.
Rule: × 10	
Input	Output
40	
600	
750	
1,000	

11.
Rule: × 100	
Input	Output
35	
100	
368	
1,000	

12.
Rule: × 1,000	
Input	Output
9	
68	
111	
100	

4-Digit Dividends

Name _____

Find the price for each item if purchased in quantity. Write >, <, or = to compare the price of each item bought in quantity to the price if it is bought separately. Use division to check.

	Teresa's Trinkets	Price in Quantity	Price Each (if bought in quantity)	>, <, or =	Price Each (if bought separately)	
1.	yo-yos	2 for $1.00			$0.63	Check
2.	kazoos	2 for $1.10			$0.55	Check
3.	bubbles	4 for $1.20			$0.35	Check
4.	bouncy balls	6 for $1.50			$0.25	Check
5.	pinwheels	4 for $5.00			$0.99	Check

Write the division equation for the phrase. Solve and label your answer.

6. 36 caramels divided among 9 friends

_____

7. 24 roses divided into 3 bouquets

_____

8. 63 baseball cards divided onto 7 pages

_____

Solve.

9. $6\overline{)6,431}$

10. $9\overline{)9,398}$

11. $18\overline{)493}$

12. $91\overline{)399}$

13. $\frac{63}{9} =$ _____

14. $56 \div 8 =$ _____

15. $8\overline{)64}$

16. $\frac{28}{4} =$ _____

Use multiplication to check the solved division problem. Circle the incorrect division problem.

1.

$$\begin{array}{r} 76 \\ 3\overline{)196} \\ -18 \\ \hline 16 \\ -16 \\ \hline 0 \end{array}$$

Check

2.

$$\begin{array}{r} 157 \\ 3\overline{)471} \\ -3 \\ \hline 17 \\ -15 \\ \hline 21 \\ -21 \\ \hline 0 \end{array}$$

Check

Solve and label your answer. Draw dollar bills and coins to illustrate.

3. Mrs. Petty gave her 3 grandchildren $18.75 to purchase 3 small craft kits. If they split the money evenly, how much money will each child get to spend?

_____

Use the rules of division to find the missing part.

4. The divisor is 7.
   The dividend is 211.
   The quotient is _____.

5. The divisor is 5.
   The dividend is _____.
   The quotient is 23.

6. The divisor is _____.
   The dividend is 233.
   The quotient is 29 r1.

Write the number in word form.

7. 0.256 _____

8. 3.213 _____

9. 5.79 _____

10. 7.04 _____

11. 0.88 _____

Solve.

12. $412 \div 4 =$ _____

13. $525 \div 5 =$ _____

14. $657 \div 9 =$ _____

15. $242 \div 2 =$ _____

16. $777 \div 7 =$ _____

17. $810 \div 9 =$ _____

Estimating

Name _____

Estimate the range.

1.  200–300

    4)823

2. 6)2,860

3. 7)194

4. 5)194

Solve.

5. 8)4,7 3 0

6. 7)6,3 0 5

7. 3)1,9 3 7

8. 5)9,1 7 2

Complete the table.

9. Mrs. Henning has 5 tables in her art room. Each table seats 5 students. Fill in the information below to show how many students will sit at each table. Clue: Mrs. Henning will completely fill a table before seating students at the next one.

Class	Number of Students	Table 1	Table 2	Table 3	Table 4	Table 5
A	17	5	5	5	2	
B	23					
C	15					
D	24					

Write a division equation to find the missing factor. Solve for *n*.

10. $3 \times n = 18$

    $n = 18 \div 3$

    $n = $ _____

11. $5 \times n = 130$

    _____

    _____

12. $n \times 7 = 343$

    _____

    _____

13. $n \times 9 = 729$

    _____

    _____

Add.

14. 453,167
    + 257,148

15. 997
    + 854

16. 5,792,163
    + 845,017

17. 75,041,638
    + 24,957,462

18. $6.93
    + 4.58

Complete the table to answer the questions.

Stephen's Bike-Riding Log			
May	Miles	Hours	Average Miles per Hour
week 1	39	3	
week 2	75	5	
week 3	56	4	
week 4	84	6	

1. Which weeks did Stephen average 14 miles per hour?

   _____

2. Which week had the lowest average miles per hour?

   _____

3. Which week did he average 15 miles per hour?

   _____

Use mental math to solve.

4. $480 \div 8 =$ _____

5. $540 \div 6 =$ _____

6. $400 \div 8 =$ _____

7. $270 \div 3 =$ _____

Draw a line through each row with incorrect output data.

8.

Rule: × 7	
Input	Output
40	280
50	350
90	640
200	1,400
600	420
700	4,900
800	5,600

9.

Rule: ÷ 9	
Input	Output
54	5
270	30
450	60
810	90
1,800	200
3,600	400
6,300	700

10.

Rule: ÷ 6	
Input	Output
54	9
120	20
480	60
1,800	300
2,400	400
3,000	600
4,200	700

Write the fraction in decimal form.

11. $\frac{7}{10} =$ _____

12. $\frac{26}{100} =$ _____

13. $\frac{4}{100} =$ _____

14. $\frac{457}{1,000} =$ _____

Write the decimal in fraction form.

15. $0.06 =$ _____

16. $0.344 =$ _____

17. $0.2154 =$ _____

18. $0.002 =$ _____

Divide.

19. $28 \div 4 =$ _____

20. $7\overline{)56}$

21. $\frac{15}{5} =$ _____

22. $\frac{81}{9} =$ _____

23. $4\overline{)28}$

24. $54 \div 6 =$ _____

25. $36 \div 6 =$ _____

26. $3\overline{)21}$

Short Form of Division

Name _____

Use the short form to solve.

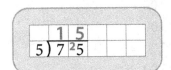

1. 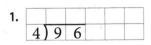 4⟌9 6

2. 6⟌3 1 7

3. 9⟌2,9 4 3

Write a division equation for the picture.

4.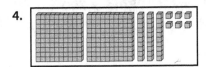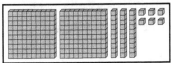

_____

5.

_____

Fill in the missing digits. Use multiplication to check.

6.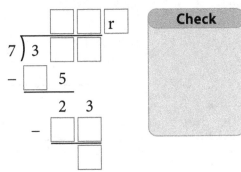

Check

7.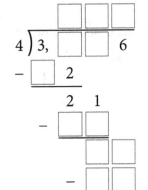

Check

Solve and label your answer.

8. Troy and Jim have a lawn business. They bought a mower for $189 and an edger for $67. They split the cost of equipment. How much did each boy spend?

9. Mrs. Peterson found ground beef on sale for $1.39 per pound. She also found it in bulk at $1.09 per pound if she purchases 3 lbs or more. How much will she save if she buys 3 pounds? 5 pounds? 10 pounds?

Find the average. Use the table to answer the question.

1. The Smiths bought tickets to the circus. They have 3 children, aged 5, 7, and 8. What is the average cost per person, including 2 tickets for the parents?

Ticket Prices	
Age	Cost
0–5	free
6–8	$10.00
9–12	$15.00
13–17	$20.00
adult	$25.00

Complete the table.

2.

Rule: + $50.00	
Input	Output
$150.00	
$275.00	
$550.00	
$925.00	

3.

Rule: – $35.00	
Input	Output
$100.00	
$275.00	
$525.00	
$1,000.00	

4.

Rule: × 700	
Input	Output
4	
7	
9	
10	

Classify the angle as *acute*, *obtuse*, *right*, or *straight*.

5.

_____

6.

_____

7.

_____

8.

_____

Write the number in standard form.

9. $(3 \times 1) + (2 \times 0.1) + (5 \times 0.01) + (6 \times 0.001) =$ _____

10. $(5 \times 1{,}000) + (7 \times 100) + (4 \times 1) =$ _____

11. $(9 \times 10{,}000) + (5 \times 100) + (3 \times 10) + (0 \times 1) =$ _____

Solve.

12. _____ $\times 7 = 56$

13. $4 \times$ _____ $= 28$

14. $9 \times$ _____ $= 18$

15. _____ $\times 6 = 36$

16. $3 \times$ _____ $= 60$

17. _____ $\times 10 = 500$

18. $70 \times$ _____ $= 7{,}000$

19. _____ $\times 1 = 1$

20. _____ $\times 7 = 49$

21. $8 \times$ _____ $= 48$

22. _____ $\times 9 = 54$

23. _____ $\times 5 = 75$

Division Review

Mark the answer. Mark *NH* if the answer is "Not Here."

1. Travis practiced for the 50-yard dash. His finishing times were 7.5 seconds, 8.3 seconds, and 6.4 seconds. What is his average?

   ○ 7.3   ○ 7.4   ○ NH

2. Peggy and Sasha baked 6 dozen cookies. They baked 9 cookies in each batch. How many batches did they bake?

   ○ 8   ○ 9   ○ NH

Use the table to find the answer.
Mark *NH* if the answer is "Not Here."

Grandpa and Eric went fishing after school three days in a row and kept a record of the number of fish each caught.

Day	Grandpa's Fish	Eric's Fish
1	4	4
2	5	2
3	3	6

3. What is the total number of fish caught in three days?

   ○ 9   ○ 25   ○ NH

4. What is the average number of fish caught by Grandpa?

   ○ 4   ○ 3   ○ NH

5. How does Eric's average number of fish compare to Grandpa's average?

   ○ greater   ○ less   ○ equal

Mark the answer. Mark *NH* if the answer is "Not Here."

6. $5 \times n = 80$
   $n = \underline{\,?\,}$

   ○ 6
   ○ 8
   ○ 9
   ○ NH

7. $5)\overline{37} = \underline{\,?\,}$

   ○ 5 r3
   ○ 6 r4
   ○ 7 r2
   ○ NH

8. $\$6.00 \div 2 = \underline{\,?\,}$

   ○ $30.00
   ○ $3.00
   ○ $0.30
   ○ NH

9. $70 \times 80 = \underline{\,?\,}$

   ○ 540
   ○ 560
   ○ 5,400
   ○ NH

10. $800 \div 4 = \underline{\,?\,}$

    ○ 20
    ○ 50
    ○ 500
    ○ NH

11. $(3 \times 10) + (5 \times 1) = \underline{\,?\,}$

    ○ 3.5
    ○ 35
    ○ 350
    ○ NH

12. The average of 2, 5, and $8 = \underline{\,?\,}$.

    ○ 2
    ○ 5
    ○ 6
    ○ NH

Solve.

13.
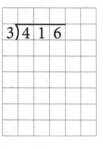
$$2\overline{)9\;7}$$

14.

$$3\overline{)4\;1\;6}$$

15.
$$5\overline{)1{,}0\;0\;2}$$

16.
$$7\overline{)5{,}9\;6\;4}$$

Write a division equation to find the missing factor.

17. $3 \times n = 12$

$n =$ _____ ÷ _____

$n =$ _____

18. $7 \times n = 91$

$n =$ _____ ÷ _____

$n =$ _____

19. $n \times 17 = 136$

$n =$ _____ ÷ _____

$n =$ _____

Write the quotient.

20. $\dfrac{54}{9} =$ _____

21. $\dfrac{81}{9} =$ _____

22. $7\overline{)49} =$ _____

23. $63 \div 9 =$ _____

Complete the table.

24.

Rule: ÷ 9	
Input	Output
9	
180	
270	
18,000	

Mark the answer.

1. If $20 \times 13 = 260$,
   and $7 \times 13 = 91$,
   then $27 \times 13 = \underline{}$.

   A. $260 - 91$ C. 260×91

   B. $260 + 91$ D. $260 \div 91$

2. If $61 \times 9 = 549$,
   and $61 \times 30 = 1{,}830$,
   then $61 \times \underline{} = 2{,}379$.

   A. 91   C. 61

   B. 70   D. 39

3. Which is *not* a name for 18?

   A. 2×9 C. $20 - 3$

   B. $10 + 6 + 2$ D. $2 \times (3 \times 3)$

4. Which is *not* a name for 40?

   A. $40 \div 10$ C. $5 \times (4 \times 2)$

   B. $(5 \times 2) \times (2 \times 2)$ D. four tens

5. Which is *not* a name for 657?

   A. 219×3 C. $7 + 600 + 50$

   B. six hundred   D. $300 + 300 + 57$
   seventy-five

6. Which number is a multiple of 10?

   A. 488   C. 456

   B. 470   D. 425

7. Which number is a multiple of 7?

   A. 42   C. 77

   B. 56   D. all of the above

8. The number 13 is a prime number.
   Which of the following is prime?

   A. 15   C. 21

   B. 17   D. none of the above

9. The number 49 is a composite number.
   Which of the following is composite?

   A. 15   C. 81

   B. 64   D. all of the above

10. $\underline{}$ rounds to 8,000.

    A. 7,346   C. 7,782

    B. 8,647   D. 8,950

11. $\underline{}$ rounds to 105,000.

    A. 104,986   C. 105,683

    B. 104,252   D. 106,900

12. $\underline{}$ rounds to 3.

    A. 2.18   C. 3.42

    B. 1.99   D. 4.01

Use rounding to estimate the answer.

13. $\begin{array}{r} 48{,}921 \\ -\ 24{,}320 \end{array}$

A. 23,601
B. 30,000
C. 73,241
D. 80,000

14. $\begin{array}{r} 591{,}722 \\ +\ 332{,}681 \end{array}$

A. 820,000
B. 900,000
C. 950,000
D. not given

15. $\begin{array}{r} 38 \\ \times\ 8 \end{array}$

A. 320
B. 420
C. 470
D. not given

16. $\begin{array}{r} 73 \\ \times\ 61 \end{array}$

A. 49
B. 480
C. 4,200
D. 3,500

Use the triangle to find the answer.

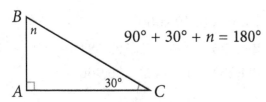

$90° + 30° + n = 180°$

17. A. ∠ABC is less than 90°.
    B. ∠ABC is greater than 90°.
    C. ∠ABC is equal to 90°.
    D. ∠ABC is a right angle.

18. A. $n = 90°$
    B. $n = 60°$
    C. $n = 180°$
    D. $n = 300°$

Use the graph to find the answer.

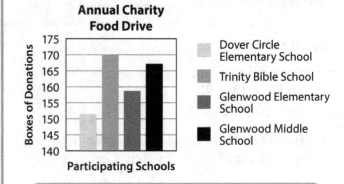

Annual Charity Food Drive

Boxes of Donations / Participating Schools

Dover Circle Elementary School
Trinity Bible School
Glenwood Elementary School
Glenwood Middle School

Stella's school collected 158 boxes of canned vegetables. Emilio's school collected 170 boxes of canned vegetables.

19. How many more boxes of vegetables did Emilio's school collect than Stella's school?
    A. 158 + 170 = 328 boxes
    B. 158 + 170 = 328 cans
    C. 170 – 158 = 12 boxes
    D. 170 – 158 = 12 cans

20. Which school does Stella attend?
    A. Dover Circle Elementary School
    D. Trinity Bible School
    C. Glenwood Elementary School
    D. Glenwood Middle School

21. Which school does Emilio attend?
    A. Dover Circle Elementary School
    B. Trinity Bible School
    C. Glenwood Elementary School
    D. Glenwood Middle School

22. Which equation shows about how many boxes were collected from these 4 schools?
    A. 150 + 170 + 160 + 165 = 645 boxes
    B. 140 + 170 + 160 + 165 = 635 boxes
    C. 170 + 170 + 170 + 170 = 680 boxes
    D. none of the above

Comparing & Ordering Fractions

Name _____

Draw a picture to show the fraction.

1. $\frac{3}{4}$ of a candy bar

2. $\frac{5}{8}$ of a pie

3. $\frac{1}{2}$ of a triangle

Draw a picture for the sentence. Write the fraction.

4. Six of the 10 jellybeans are red.

_____

5. Two of the 6 boxes are full.

_____

6. Four of the 8 apples are yellow.

_____

Write the numerators that make each fraction equal to 1.

7. $\overline{7}$ $\overline{4}$ $\overline{5}$ $\overline{3}$ $\overline{12}$

Write the numerators that make each fraction equal to $\frac{1}{2}$.

8. $\overline{4}$ $\overline{6}$ $\overline{8}$ $\overline{10}$ $\overline{12}$

Complete the number line.

9.

10.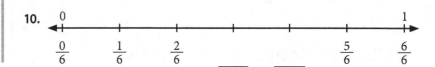

Write >, <, or = to compare.

11. $\frac{1}{7}$ ◯ $\frac{1}{6}$

12. $\frac{1}{2}$ ◯ $\frac{4}{6}$

13. $\frac{3}{4}$ ◯ $\frac{3}{6}$

14. $\frac{1}{2}$ ◯ $\frac{2}{4}$

15. $\frac{1}{3}$ ◯ $\frac{3}{7}$

16. $\frac{2}{8}$ ◯ $\frac{1}{4}$

17. $\frac{1}{3}$ ◯ $\frac{1}{5}$

18. $\frac{5}{8}$ ◯ $\frac{2}{4}$

19. $\frac{8}{8}$ ◯ $\frac{7}{8}$

Write the fractions from _least_ to _greatest_.

20. $\frac{3}{6}$ $\frac{1}{3}$ $\frac{5}{6}$

_____ _____ _____

21. $\frac{1}{2}$ $\frac{12}{12}$ $\frac{3}{12}$

_____ _____ _____

22. $\frac{1}{2}$ $\frac{1}{4}$ $\frac{1}{5}$

_____ _____ _____

Multiply.

1.	2.	3.	4.	5.
247 × 63	158 × 21	603 × 19	111 × 86	409 ×167

Divide.

6. 3)5 8 6

7. 7)2 3 9

8. 9)7 1 4

Use the graph to answer the questions.

Greg is training for a ski-jumping competition. The bar graph shows the number of hours he trained each day.

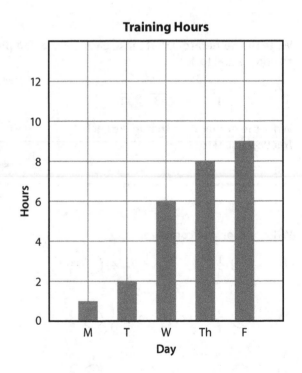

Training Hours

9. How long did Greg train on Monday?

_____

10. What is the difference between training hours Tuesday and Wednesday?

_____

11. What day did he train for 8 hours?

_____

12. What is the total hours he trained?

_____

Write the fractions from *least* to *greatest*.

13.	14.	15.
$\frac{1}{2}$ $\frac{3}{8}$ $\frac{7}{8}$	$\frac{1}{2}$ $\frac{3}{4}$ $\frac{1}{4}$	$\frac{2}{6}$ $\frac{5}{6}$ $\frac{2}{3}$

____ ____ ____   ____ ____ ____   ____ ____ ____

Renaming Fractions

Name _____

Repartition the figure to make an equivalent fraction in higher terms. Solve.

1.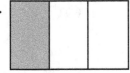

$$\frac{1}{3} = \frac{}{6}$$

2.

$$\frac{3}{4} = \frac{}{8}$$

3.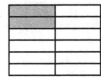

$$\frac{2}{12} = \frac{}{24}$$

4.

$$\frac{2}{4} = \frac{}{8}$$

Multiply by a name for 1 to make an equivalent fraction in higher terms.

5. $\frac{2}{7} = \frac{n}{14}$

$$\frac{2 \times}{7 \times} = \frac{}{14}$$

6. $\frac{3}{5} = \frac{n}{15}$

$$\frac{3 \times}{5 \times} = \frac{}{15}$$

7. $\frac{4}{6} = \frac{n}{24}$

$$\frac{4 \times}{6 \times} = \frac{}{24}$$

Divide the figure into the number of parts that represent an equivalent fraction in lower terms. Write the equivalent fraction in lower terms.

8.

$$\frac{3}{6} = \frac{}{2}$$

9.

$$\frac{3}{9} = \frac{}{3}$$

10.

$$\frac{6}{8} = \frac{}{4}$$

Draw a picture. Answer the questions and label your answers.

11. Elisabeth had 9 small boxes of raisins. She ate the raisins from one-third of the boxes. How many boxes of raisins did she eat? _____
How many boxes remain? _____

Solve. Rename the answer to lower terms.

12. Four of Mrs. Moore's 20 math students are competing in the mental math contest. What fraction of the class is competing? _____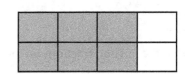

Circle the smallest fraction.

13. $\frac{1}{2}$ $\frac{1}{4}$ $\frac{1}{10}$ 14. $\frac{3}{4}$ $\frac{3}{2}$ $\frac{3}{1}$

15. $\frac{1}{4}$ $\frac{2}{8}$ $\frac{1}{16}$ 16. $\frac{4}{5}$ $\frac{3}{5}$ $\frac{4}{10}$

Circle the largest fraction.

17. $\frac{2}{8}$ $\frac{4}{8}$ $\frac{4}{4}$ 18. $\frac{6}{8}$ $\frac{5}{8}$ $\frac{4}{8}$

19. $\frac{2}{10}$ $\frac{8}{10}$ $\frac{1}{2}$ 20. $\frac{3}{5}$ $\frac{4}{8}$ $\frac{2}{4}$

Write the fraction for the part that is shaded.

1. ____

2. 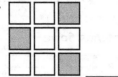 ____

3. 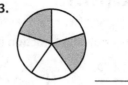 ____

4.  ____

Draw a picture for the sentence.

5. $\frac{1}{7}$ of the squares are shaded.

6. $\frac{2}{4}$ of the circle is shaded.

7. The circle is divided into 4 equal parts.

8. $\frac{1}{6}$ of the rectangle is shaded.

Write the missing fractions on the number line.

9.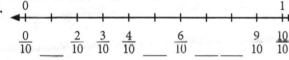

$\frac{0}{10}$ ___ $\frac{2}{10}$ $\frac{3}{10}$ $\frac{4}{10}$ ___ $\frac{6}{10}$ ___ ___ $\frac{9}{10}$ $\frac{10}{10}$

10.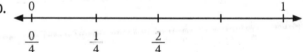

$\frac{0}{4}$ $\frac{1}{4}$ $\frac{2}{4}$ ___ ___

Circle the fractions that are equal to 1.

11. $\frac{2}{5}$ $\frac{4}{4}$ $\frac{7}{8}$ $\frac{3}{3}$ $\frac{9}{9}$ $\frac{1}{6}$

Circle the fractions that are equal to $\frac{1}{4}$.

12. $\frac{2}{3}$ $\frac{3}{12}$ $\frac{5}{20}$ $\frac{9}{10}$

Name the part of the circle. Answer the questions.

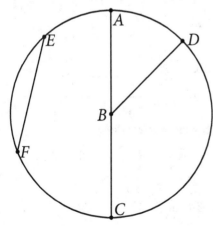

13. radius _____

14. diameter _____

15. chord _____

16. What is a chord? _____

_____

17. How is a radius named? _____

_____

18. How is a circle named? _____

Improper Fractions & Mixed Numbers

Name _____

Draw a picture. Answer the questions and label your answer.

1. Hannah is putting pictures in her photo album. She has 23 pictures. Six pictures will fit on each page. How many pages will Hannah need for her pictures?

   _____

2. Daniel made Rice Krispies® treats for his class. He cut 2 pans of treats into 12 pieces each. His class ate $1\frac{5}{6}$ of the pans. How many treats were left?

   _____

3. The choir director arranged his group of 42 members evenly onto 6 risers. What improper fraction illustrates this number? _____ How many members were on each row?

   _____

Draw a picture for the sentence. Write the improper fraction as a mixed number or whole number.

4. Show $\frac{7}{3}$ using 3 squares.

   $\frac{7}{3} = $

5. Show $\frac{9}{5}$ using 2 rectangles.

   $\frac{9}{5} = $

6. Show $\frac{4}{3}$ using squares.

   $\frac{4}{3} = $

7. Show $\frac{6}{2}$ using rectangles.

   $\frac{6}{2} = $

Write an addition equation to rename the improper fraction or the mixed number.

8. $2\frac{1}{2} = \frac{2}{2} + \frac{2}{2} + \frac{1}{2}$

9. $\frac{16}{3} =$

10. $7\frac{1}{2} =$

11. $4\frac{3}{5} =$

12. $\frac{11}{2} =$

13. $3\frac{2}{3} =$

Write the missing fractions on the number line.

1.
$$\frac{0}{5} \qquad \underline{} \qquad \frac{2}{5} \qquad \frac{3}{5} \qquad \underline{} \qquad \frac{5}{5}$$

2.
$$\frac{0}{14} \qquad \frac{2}{14} \qquad \underline{} \qquad \frac{7}{14} \qquad \underline{} \qquad \frac{12}{14} \qquad \frac{14}{14}$$

Multiply by a name for 1 to make an equivalent fraction in higher terms. Solve for n.

3. $\dfrac{3}{4} = \dfrac{n}{24}$

$$\frac{3 \times }{4 \times } = \frac{}{24}$$

4. $\dfrac{2}{5} = \dfrac{n}{25}$

$$\frac{2 \times }{5 \times } = \frac{}{25}$$

5. $\dfrac{6}{7} = \dfrac{n}{21}$

$$\frac{6 \times }{7 \times } = \frac{}{21}$$

Draw a picture.

6. Shade $\frac{1}{2}$ of a circle. Repartition to show sixths.

7. Shade $\frac{4}{5}$ of a rectangle. Repartition to show tenths.

Write >, <, or = to compare.

8. $\dfrac{1}{4} \bigcirc \dfrac{3}{3}$

9. $\dfrac{1}{2} \bigcirc \dfrac{2}{4}$

10. $\dfrac{7}{4} \bigcirc \dfrac{1}{4}$

11. $\dfrac{16}{12} \bigcirc \dfrac{1}{2}$

12. $\dfrac{3}{7} \bigcirc \dfrac{2}{7}$

13. $\dfrac{3}{4} \bigcirc \dfrac{2}{5}$

14. $\dfrac{5}{10} \bigcirc \dfrac{3}{15}$

15. $\dfrac{3}{8} \bigcirc \dfrac{3}{6}$

16. $\dfrac{6}{12} \bigcirc \dfrac{1}{2}$

17. $\dfrac{2}{4} \bigcirc \dfrac{4}{2}$

18. $\dfrac{4}{3} \bigcirc \dfrac{6}{6}$

19. $1\dfrac{1}{8} \bigcirc \dfrac{8}{4}$

Comparing Mixed Numbers   Name _____

Write the fraction that represents the paint colors used by each class.

1. The art teacher noticed that her students used up
   certain colors from their paint sets more quickly than
   others. Her third grade class used up 7 of the 8 colors,
   her fourth grade class used up 6 of the 8 colors, and
   her fifth grade class used up 5 of the 8 colors.

   ____ ____ ____
   third fourth fifth

Solve and label your answer.

> The zookeeper feeds fish to the penguins from a pail that holds 12 fish.

2. At the morning feeding, the penguins ate 33 fish.
   How many full pails of fish did they eat?   _____

   What fractional part of the next pail did they eat?   _____

   Write a mixed number to tell how many pails of
   fish were eaten.   _____

3. At the evening feeding, the penguins ate 41 fish.
   Write a mixed number to represent how many
   pails of fish were eaten.   _____

Round to the nearest whole number.

4.  _____

5. 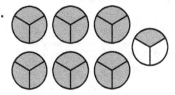 _____

6. $2\frac{6}{7}$ _____   7. $3\frac{1}{5}$ _____   8. $5\frac{4}{5}$ _____

Write >, <, or = to compare.

9. $\frac{9}{4}$ ◯ $2\frac{2}{4}$ 10. $1\frac{5}{6}$ ◯ $\frac{13}{6}$ 11. $\frac{17}{9}$ ◯ $1\frac{1}{9}$ 12. $3\frac{1}{7}$ ◯ $\frac{23}{7}$

13. five and six-tenths ◯ six and five-tenths   14. two and one-half ◯ $2\frac{1}{3}$

Write the mixed numbers from *least* to *greatest*.

15. | $4\frac{2}{9}$ | $2\frac{1}{6}$ | $2\frac{2}{3}$ |

    ____ ____ ____

16. | $2\frac{1}{2}$ | $2\frac{3}{5}$ | $2\frac{1}{6}$ |

    ____ ____ ____

Complete the table. Write an improper fraction and a mixed number to represent the amount of pizza the class ate. Answer the questions.

1.

Pizza Order			
Topping	Quantity Eaten	Improper Fraction	Mixed Number
pepperoni	(3 circles in eighths, shaded: 2 full + 6/8)	$\frac{22}{8}$	$2\frac{6}{8}$
sausage	(2 circles in eighths)		
cheese	(3 circles in tenths)		
veggie	(2 circles in eighths)		

2. How many pieces of pepperoni pizza were left? _____

3. How many pieces of cheese pizza were left? _____

Write an improper fraction and a mixed number.

4.

_____ _____
improper   mixed
fraction   number

5. $\frac{1}{3} + \frac{1}{3} + \frac{1}{3} + \frac{1}{3} + \frac{1}{3}$

_____ _____
improper   mixed
fraction   number

6.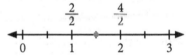

_____ _____
improper   mixed
fraction   number

Write a mathematical expression for the phrase.

7. nine less than 24 _____

8. five in each of 9 sets _____

9. three sets of 12 _____

10. seventeen and thirteen more _____

11. one hundred minus 21 _____

12. twelve and nine more _____

Solve.

13. $4 + 9 + 3 + 6 =$ _____

14. $(42 \div 7) + 31 + 9 =$ _____

15. $\frac{1}{5} + \frac{6}{5} + \frac{3}{5} =$ _____

16. $(3 \times 1) + (4 \times 7) =$ _____

17. $(16 \div 4) \times (5 \times 1) =$ _____

18. $2\frac{1}{3} + 3\frac{2}{3} =$ _____

Common Factors

Name _____

Answer the riddle.

1. I am a prime number between 6 and 11. _____

2. I am a composite number with the factors 1, 2, 7, 14. _____

3. We are the factors for 9. _____

4. We are the common factors for 21 and 63. _____

Use the Venn diagram to complete the statement.

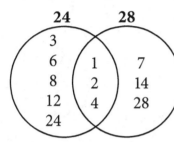

5. List the factors.

   24: _____

   28: _____

   The common factors of 24 and 28 are _____, _____, and _____.

List the factors. Complete the Venn diagram.

6. 32: _____, _____, _____, _____, _____, _____

   36: _____, _____, _____, _____, _____, _____, _____, _____, _____

   The common factors of 32 and 36 are _____, _____, and _____.

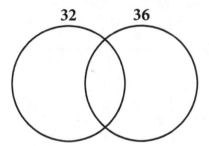

Circle the fractions that are in lowest terms. Rename the remaining fractions to lowest terms.

7. $\frac{4}{8}$ $\frac{2}{3}$ $\frac{5}{7}$ $\frac{3}{4}$ $\frac{2}{6}$

Write the fraction. Rename the fraction in lowest terms.

8. 4 brownies left from a pan of 24 _____

9. 13 boys in a Sunday school class of 26 _____

10. 6 eggs left in a carton of 18 _____

11. 5 miles left in a drive of 25 miles _____

12. 10 days left in a 30-day month _____

Write >, <, or = to compare.

1. $3\frac{1}{3}$ ◯ $2\frac{2}{3}$

2. $1\frac{2}{4}$ ◯ $1\frac{1}{2}$

3. $\frac{6}{9}$ ◯ $\frac{2}{3}$

4. 7 ◯ $\frac{36}{6}$

5. $\frac{11}{5}$ ◯ $2\frac{1}{5}$

6. $2\frac{5}{8}$ ◯ $1\frac{6}{8}$

7. $\frac{7}{6}$ ◯ $1\frac{1}{3}$

8. $\frac{7}{8}$ ◯ $\frac{3}{4}$

Rename the improper fraction as a mixed number.

	improper fraction	mixed number
9.	$\frac{10}{3}$	
10.	$\frac{12}{8}$	
11.	$\frac{7}{2}$	
12.	$\frac{7}{3}$	

Rename the mixed number as an improper fraction.

	improper fraction	mixed number
13.		$3\frac{1}{6}$
14.		$4\frac{3}{8}$
15.		$2\frac{5}{6}$
16.		$1\frac{10}{11}$

Classify the angle as *acute*, *obtuse*, or *right*.

17.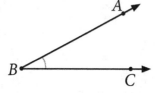

_____

18.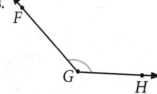

_____

19.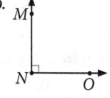

_____

20. Mr. Simmons cut mosaic tiles on a 45° angle. _____

21. Ridge Road and Groce Road meet at a 120° angle. _____

22. The angle is made by the corner of a room. _____

Complete the equation with +, −, ×, or ÷.

23. 3 ◯ $4 = 12$

24. 18 ◯ $9 = 2$

25. 4 ◯ $4 = 16$

26. 3 ◯ $1 = 4$

27. 8 ◯ $9 = 17$

28. 9 ◯ $9 = 18$

29. 4 ◯ $4 = 8$

30. 3 ◯ $1 = 3$

31. 24 ◯ $2 = 12$

32. 9 ◯ $9 = 81$

33. 16 ◯ $8 = 2$

34. 3 ◯ $3 = 1$

35. 6 ◯ $5 = 30$

36. 18 ◯ $9 = 9$

37. 2 ◯ $8 = 16$

38. 3 ◯ $0 = 0$

Lowest Terms

Name _____

List the factors of the numerator and denominator. Use the greatest common factor to repartition the picture. Write the equivalent fraction in lowest terms.

1. $\frac{4}{8}$

4: 1, 2, 4 _____

8: 1, 2, 4, 8 _____

$\frac{4 \div 4}{8 \div 4} = \frac{\ }{\ }$

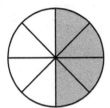

2. $\frac{6}{9}$

6: _____

9: _____

$\frac{6 \div}{9 \div} = \frac{\ }{\ }$

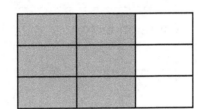

3. $\frac{10}{12}$

10: _____

12: _____

$\frac{10 \div}{12 \div} = \frac{\ }{\ }$

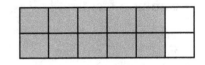

Rename the fraction in lowest terms.

4. $\frac{15}{20} = \frac{n}{4}$ _____

5. $\frac{6}{12} = \frac{n}{4}$ _____

6. $\frac{10}{22} = \frac{n}{11}$ _____

7. $\frac{14}{21} = \frac{n}{3}$ _____

8. $\frac{15}{21} = \frac{n}{7}$ _____

9. $\frac{6}{18} = \frac{n}{6}$ _____

Circle the fractions that are in lowest terms. Rename the remaining fractions to lowest terms.

10. $\frac{2}{6}$ $\frac{1}{3}$

11. $\frac{4}{5}$

12. $\frac{5}{9}$

13. $\frac{7}{21}$

14. $\frac{4}{21}$

15. $\frac{12}{13}$

16. $\frac{16}{24}$

17. $\frac{10}{24}$

Write the fraction. Rename the fraction in lowest terms.

18. Mrs. Boyce has 6 puppies. Two are brown and four are white. Write the number of brown and white puppies as fractions and rename to lowest terms.

brown: _____   white: _____

19. A comic book has 28 pages with 11 in color. Write the number of color pages as a fractional part of the book.

color pages: _____

Write the fraction. Rename the fraction in lowest terms.

1.
_____

2.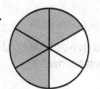
_____

Write the factors of the numerator and denominator. Find the common factors. Divide by the greatest common factor listed to rename the fraction.

3. $\frac{5}{20}$

5: __1, 5_____

20: __1, 2, 4, 5, 10, 20__

common factors: __1, 5____

$\frac{5 \div}{20 \div} = \frac{}{}$

4. $\frac{4}{12}$

4: _____

12: _____

common factors: _____

$\frac{4 \div}{12 \div} = \frac{}{}$

5. $\frac{9}{15}$

9: _____

15: _____

common factors: _____

$\frac{9 \div}{15 \div} = \frac{}{}$

Complete the Venn diagram to find the factors.

6. **36   42**

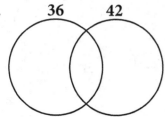

List the factors.

36: _____

42: _____

List the common factors of 36 and 42.

_____

7. **16   28**

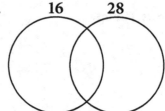

List the factors.

16: _____

28: _____

List the common factors of 16 and 28.

_____

Write the fraction. Rename the fraction in lower terms.

8. 12 oz used up from a 16-oz bottle _____

9. 9 broken crayons from a pack of 24 _____

10. 3 eggs used out of a dozen _____

Multiply.

11. $\begin{array}{r} 17 \\ \times\ 4 \\ \hline \end{array}$

12. $\begin{array}{r} 63 \\ \times 11 \\ \hline \end{array}$

13. $\begin{array}{r} 219 \\ \times\ \ 6 \\ \hline \end{array}$

14. $\begin{array}{r} 467 \\ \times\ 50 \\ \hline \end{array}$

15. $\begin{array}{r} \$10.00 \\ \times\ \ \ \ \ 9 \\ \hline \end{array}$

16. $\begin{array}{r} 809 \\ \times\ 12 \\ \hline \end{array}$

More Lowest Terms

Name _____

List the factors of the numerator and denominator. Find the greatest common factor (GCF).
Use the greatest common factor to write the equivalent fraction in lowest terms.

1. $\frac{6}{12}$

    6: _1, 2, 3, 6_

    12: _1, 2, 3, 4, 6, 12_

    GCF: _6_

    $\frac{6 \div 6}{12 \div 6} = $ ⎯

2. $\frac{5}{35}$

    5: _____

    35: _____

    GCF: _____

    $\frac{5 \div}{35 \div} = $ ⎯

3. $\frac{18}{24}$

    18: _____

    24: _____

    GCF: _____

    $\frac{18 \div}{24 \div} = $ ⎯

Rename the mixed number as an improper fraction.

4.

improper fraction	mixed number
	$2\frac{3}{8}$
	$4\frac{5}{6}$
	$1\frac{11}{13}$
	$7\frac{9}{12}$
	$5\frac{1}{7}$

Rename the improper fraction as a mixed number.

5.

improper fraction	mixed number
$\frac{20}{9}$	
$\frac{22}{3}$	
$\frac{5}{4}$	
$\frac{17}{4}$	
$\frac{7}{2}$	

List the factors for the numbers. Find the greatest common factor (GCF).

6. 15: _____

    25: _____

    GCF: _____

7. 18: _____

    24: _____

    GCF: _____

Make a factor tree. Follow the steps to find the greatest common factor (GCF).

8.   28   36

List the prime factors of 28 and 36.

28: _____

36: _____

Multiply the common factors.

_____ × _____ = _____

The GCF of 28 and 36 is _____.

Write the fraction. Rename the fraction in lowest terms.

9. The spelling list had 25 words.
   Fifteen of the words were adjectives. _____

10. The parking lot holds
    40 cars. 28 spaces are full. _____

Write >, <, or = to compare.

1. $\frac{3}{6}$ ◯ $\frac{2}{3}$

2. $2\frac{1}{4}$ ◯ $\frac{7}{4}$

3. $\frac{25}{6}$ ◯ $3\frac{5}{6}$

4. $\frac{11}{8}$ ◯ $1\frac{4}{8}$

5. $\frac{3}{8}$ ◯ $\frac{1}{2}$

6. 1 ◯ $\frac{4}{3}$

7. $\frac{6}{3}$ ◯ $\frac{9}{5}$

8. $\frac{8}{8}$ ◯ 1

Write the missing numerator.

9. $\frac{1}{6} = \frac{}{12}$

10. $1\frac{1}{3} = \frac{}{3}$

11. $\frac{15}{20} = \frac{}{4}$

12. $3\frac{}{2} = \frac{7}{2}$

13. $3 = \frac{}{5}$

14. $\frac{16}{15} = 1\frac{}{15}$

15. $\frac{}{9} = 3$

16. $1\frac{}{5} = \frac{8}{5}$

Draw a picture for the sentence.

17. $1\frac{2}{3}$ squares are shaded.

18. $\frac{3}{4}$ of a circle is shaded.

19. $3\frac{4}{7}$ rectangles are shaded.

20. $\frac{9}{4}$ squares are shaded.

Solve.

21.
```
  2,□26
+ 4,34□
───────
  □,4□8
```

22.
```
  1,9□4
-  □5□
───────
  1,753
```

23.
```
  $ 2 1 7 . 3 4
-       2 . □ □
──────────────
  $ 2 1 □ . 9 3
```

24.
```
  4,50□
+ 2,873
───────
  □,□74
```

25.
```
  16
×  9
```

26.
```
  47
×  8
```

27. $8\overline{)72}$

28. $7\overline{)56}$

29. $9\overline{)54}$

Fraction Review

Name _____

Mark the answer.
Mark *NH* if the answer is "Not Here."

1. $1 =$
   - ○ $\frac{7}{6}$
   - ○ $\frac{7}{7}$
   - ○ $\frac{6}{7}$
   - ○ NH

2. $\frac{1}{2} =$
   - ○ $\frac{3}{5}$
   - ○ $\frac{3}{6}$
   - ○ $\frac{3}{1}$
   - ○ NH

3. $\frac{11}{7} =$
   - ○ $4\frac{1}{7}$
   - ○ $7\frac{4}{7}$
   - ○ $1\frac{1}{7}$
   - ○ NH

4. 7
   - ○ prime
   - ○ composite

5. 13
   - ○ prime
   - ○ composite

6. 14
   - ○ prime
   - ○ composite

7.
   - ○ $\frac{5}{5}$ is shaded.
   - ○ $\frac{5}{10}$ is shaded.
   - ○ $\frac{5}{11}$ is shaded.
   - ○ NH

8.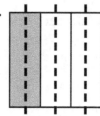
   - ○ $\frac{1}{3} = \frac{1}{6}$
   - ○ $\frac{1}{3} = 2$
   - ○ $\frac{1}{3} = \frac{2}{6}$
   - ○ NH

9. $\frac{1}{2}$ ○ $\frac{6}{8}$
   - ○ >   ○ <   ○ =

10. $\frac{4}{7}$ ○ $\frac{12}{21}$
    - ○ >   ○ <   ○ =

11. $1\frac{7}{8}$ ○ $\frac{15}{16}$
    - ○ >   ○ <   ○ =

12.
    - ○ $\frac{3}{4}$
    - ○ $\frac{6}{8}$
    - ○ $\frac{8}{8}$
    - ○ NH

    $\frac{2}{4} = \frac{?}{\underline{}}$

13.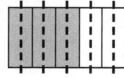
    - ○ $\frac{4}{5}$
    - ○ $\frac{5}{5}$
    - ○ $\frac{6}{10}$
    - ○ NH

    $\frac{3}{5} = \frac{?}{\underline{}}$

14.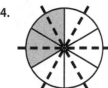
    - ○ $\frac{4}{12}$
    - ○ $\frac{6}{12}$
    - ○ $\frac{2}{3}$
    - ○ NH

    $\frac{2}{6} = \frac{?}{\underline{}}$

Mark the answer.
Mark *NH* if the answer is "Not Here."

15.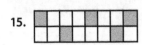
- ○ $\frac{5}{6}$ is shaded.
- ○ $\frac{5}{11}$ is shaded.
- ○ $\frac{5}{12}$ is shaded.
- ○ NH

16.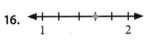
- ○ $1\frac{2}{5}$ is marked.
- ○ $\frac{3}{5}$ is marked.
- ○ $1\frac{3}{5}$ is marked.
- ○ NH

17. $\frac{13}{14}$
- ○ 1 and $\frac{1}{14}$ parts have been selected.
- ○ another name for $\frac{6}{7}$
- ○ The whole has been divided into 13 parts.
- ○ NH

Complete the factor tree.

18.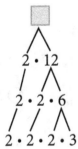
- ○ 6
- ○ 12
- ○ 24
- ○ NH

19.
- ○ 14
- ○ 21
- ○ 31
- ○ NH

Solve.

20. $\frac{4}{6} = \frac{n}{24}$

    $n = ?$
- ○ 4
- ○ 16
- ○ 1
- ○ NH

21. $\frac{2}{5} = \frac{n}{20}$

    $n = ?$
- ○ 5
- ○ 10
- ○ 8
- ○ NH

Write the fractions from *least* to *greatest*.

22.
| $\frac{2}{3}$ | $\frac{5}{3}$ | $1\frac{1}{3}$ |

_____   _____   _____

23.
| $\frac{1}{2}$ | $\frac{1}{4}$ | $\frac{1}{3}$ |

_____   _____   _____

Write an improper fraction and a mixed number for the picture.

24.

_____   _____
improper fraction   mixed number

25.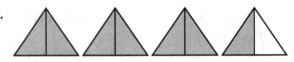

_____   _____
improper fraction   mixed number

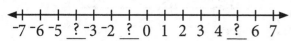

Name _____

Mark the answer.

$$\overset{\longleftarrow}{\underset{-7\ -6\ -5\ \underline{?}\ -3\ -2\ \underline{?}\ 0\ 1\ 2\ 3\ 4\ \underline{?}\ 6\ 7}{\vert\ \vert\ \vert\ \vert\ \vert\ \vert\ \vert\ \vert\ \vert\ \vert\ \vert\ \vert\ \vert\ \vert\ \vert}}\overset{\longrightarrow}{}$$

1. Which numbers are missing from the number line?

   A. ‾6, ‾1, 5
   B. ‾4, ‾1, ‾5
   C. 6, 1, 5
   D. ‾4, ‾1, 5

2. 2 ◯ ‾2

   A. >
   B. <
   C. =

3. ‾6 ◯ 3

   A. >
   B. <
   C. =

4. 4 ◯ ‾7

   A. >
   B. <
   C. =

5. Choose the numbers that are listed *least* to *greatest*.

   A. ‾1, ‾3, ‾5, ‾7
   B. ‾6, ‾2, 4, 6
   C. 7, 5, 0, ‾1, ‾3
   D. 7, 5, 0, ‾3, ‾1

6. 6 × 6 ◯ 30 + 7

   A. >
   B. <
   C. =

7. 36 ÷ 4 ◯ 25 − 16

   A. >
   B. <
   C. =

8. 42 + 16 ◯ 7 × 8

   A. >
   B. <
   C. =

9. 13 + 57 ◯ 62 − 12

   A. >
   B. <
   C. =

10. 28 ÷ 7 ◯ 27 ÷ 3

    A. >
    B. <
    C. =

11. 9 × 4 ◯ 12 × 3

    A. >
    B. <
    C. =

12. 10 × 6 ◯ 74 − 18

    A. >
    B. <
    C. =

13. 17 − 6 ◯ 20 ÷ 2

    A. >
    B. <
    C. =

Mark the answer.

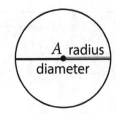

14. The diameter of circle *A* is 44 cm. What is the radius?

    A. 42 cm

    B. 88 cm

    C. 22 cm

    D. 11 cm

15. The radius of circle *A* is 12 units. What is the diameter?

    A. 10 units

    B. 6 units

    C. 4 units

    D. 24 units

16. ∠*LTM* is __?__.

    A. acute

    B. obtuse

    C. right

17. ∠*PRN* is __?__.

    A. acute

    B. obtuse

    C. right

18. ∠*IZX* is __?__.

    A. acute

    B. obtuse

    C. right

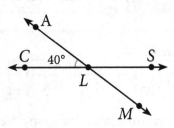

19. ∠*ALS* = __?__

    A. 40°

    B. 60°

    C. 115°

    D. 140°

20. ∠*CLS* is __?__.

    A. acute

    B. straight

    C. obtuse

    D. right

21. $\overline{AL}$ is part of __?__.

    A. $\overleftrightarrow{AM}$

    B. ∠*SLM*

    C. $\overleftrightarrow{CS}$

    D. $\overrightarrow{LC}$

22. $\overrightarrow{LC}$ is part of __?__.

    A. ∠*SLM*

    B. $\overrightarrow{LM}$

    C. $\overleftrightarrow{AM}$

    D. $\overleftrightarrow{CS}$

Multiples of 10

Name _____

Write the basic fact. Use mental math to solve.

1. (___ ÷ ___)

    20)‾6‾0‾0‾

2. (___ ÷ ___)

    50)‾3‾0‾0‾

3. (___ ÷ ___)

    40)‾1‾6‾0‾

4. (___ ÷ ___)

    90)‾8‾,‾1‾0‾0‾

5. (___ ÷ ___)

    30)‾2‾,‾7‾0‾0‾

Complete a division problem to continue the pattern.

6.
$$\frac{2}{60)120} \qquad \frac{20}{60)1,200} \qquad \frac{200}{60)12,000} \qquad 60)\overline{}$$

7.
$$\frac{9}{80)720} \qquad \frac{90}{80)7,200} \qquad \frac{900}{80)72,000} \qquad 80)\overline{}$$

8.
$$\frac{8}{30)240} \qquad \frac{80}{30)2,400} \qquad \frac{800}{30)24,000} \qquad 30)\overline{}$$

Estimate the quotient. Solve.

9. | Estimate |

    30)‾1‾9‾0‾

10. | Estimate |

    60)‾4‾5‾0‾

11. | Estimate |

    40)‾1‾7‾0‾

Solve.

12.

13.

14.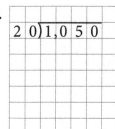

15.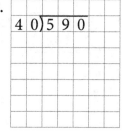

Use mental math to solve. Label your answer.

16. Calvary Church collected 280 blankets to give to the local homeless shelter. They packed the blankets in 40 boxes, distributing the blankets evenly. How many blankets were in each box?

17. Mr. Fisher purchased 2,500 new business cards. They arrived evenly divided among 10 boxes. How many cards were in each box?

18. Evan's new cell phone plan allows him to text up to 150 times per month. On a thirty-day month, how many texts can he send per day on average?

Write the fraction in decimal form.

1. $\frac{6}{10} =$ _____

2. $\frac{14}{100} =$ _____

3. $\frac{56}{100} =$ _____

4. $\frac{399}{1,000} =$ _____

Write > or < to compare.

5. $6 \bigcirc {}^-6$

6. ${}^-8 \bigcirc {}^-2$

7. $4 \bigcirc {}^-12$

8. ${}^-9 \bigcirc 4$

9. $1 \bigcirc {}^-3$

Write the numbers from *least* to *greatest*.

10.
⁻5	1	⁻3	3

____ ____ ____ ____

11.
9	⁻15	⁻6	18

____ ____ ____ ____

Solve to complete the model.

12. $47 - n = 24$

$n = \underline{47 -}$

$n = $ _____

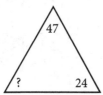

13. $n + 14 = 75$

$n = $ _____

$n = $ _____

14. $22 + n = 43$

$n = \underline{43 -}$

$n = $ _____

43	
22	

15. $66 - n = 31$

$n - $ _____

$n = $ _____

66	
	31

Divide.

16. $10\overline{)40}$

17. $8\overline{)24}$

18. $2\overline{)8}$

19. $20\overline{)80}$

20. $8\overline{)64}$

21. $40\overline{)120}$

22. $5\overline{)25}$

23. $50\overline{)250}$

24. $9\overline{)81}$

25. $30\overline{)120}$

26. $8\overline{)40}$

27. $80\overline{)400}$

28. $6\overline{)36}$

29. $60\overline{)360}$

30. $9\overline{)54}$

1-Digit Quotients

Name _____

Complete the chart.

Ann kept a journal of the calories from fruit she had eaten in 1 week. How many servings of each fruit did she eat?

Solve

1.

Fruit	Calories per Serving	Calories in 1 Week	Number of Fruit Servings
peach	35	245	
apple	44	396	
kiwi	34	204	
pear	45	315	
pineapple	50	200	

Solve and label your answer.

2. Twin Valley Christian Elementary School has 153 students. Each classroom has an average of 17 students. How many elementary classrooms does Twin Valley have?

3. The school secretary purchased a supply of 850 paper clips. Paper clips are boxed with 50 in a box. How many boxes did she purchase?

4. Mrs. Hutton has 39 violin students with 13 students per class. How many classes does she teach?

5. Mr. Daniels has 66 bookmarks to give to his 22 students. How many bookmarks can each student receive?

Solve. Use multiplication to check.

6.

$$17\overline{)91}$$

Check

7.

$$16\overline{)68}$$

Check

8.

$$90\overline{)507}$$

Check

Solve and label your answer.

1. Calvary Baptist Church took a special offering to buy Bibles and to help missionaries in Brazil. In Sunday school $424.98 was collected. During the morning service $563.02 was collected. At the evening service $349.22 was collected. How much money was collected altogether?

Solve

2. The church used $250 to purchase Portuguese Bibles to send to the missionaries. How much money was left?

Solve

Use mental math to solve.

3. $10)\overline{4,000}$

4. $70)\overline{560}$

5. $20)\overline{240}$

6. $30)\overline{6,000}$

7. $80)\overline{640}$

8. $40)\overline{80}$

9. $60)\overline{120}$

10. $90)\overline{180}$

11. $50)\overline{2,500}$

12. $10)\overline{30}$

Use mental math to solve. Label your answer.

13. Mrs. Smith's Sunday school class bought 1,200 tracts to send to missionaries. The tracts are equally divided among 20 designs. How many tracts are there of each design?

Divide.

14. $6)\overline{54}$

15. $5)\overline{40}$

16. $9)\overline{72}$

17. $3)\overline{27}$

18. $4)\overline{28}$

19. $42 \div 7 = $ ___

20. $35 \div 5 = $ ___

21. $56 \div 8 = $ ___

22. $45 \div 9 = $ ___

23. $70 \div 10 = $ ___

Adjusting the Quotient

Name _____

Complete the problem. Mark the answer.

1.
   ○ Adjust the quotient up.
   ○ Adjust the quotient down.
   ○ The quotient is correct.

2.
   ○ Adjust the quotient up.
   ○ Adjust the quotient down.
   ○ The quotient is correct.

3.
   ○ Adjust the quotient up.
   ○ Adjust the quotient down.
   ○ The quotient is correct.

4.
   ○ Adjust the quotient up.
   ○ Adjust the quotient down.
   ○ The quotient is correct.

Round the divisor. Estimate the quotient.

5.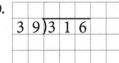
$21\overline{)76}$

6.
$27\overline{)104}$

7.
$46\overline{)316}$

8.
$19\overline{)590}$

Solve. Use multiplication to check.

9.
$39\overline{)316}$

Check

10.
$77\overline{)604}$

Check

11.
$24\overline{)236}$

Check

Solve and label your answer.

12. Mr. Matthews spent 104 hours building a detailed model battleship. If he spent 8 hours a day working on the model, how many days did it take him to finish it?

13. Hutton's Shoe Shop has 408 pairs of shoes in stock. If they sell an average of 51 pairs per week, how many weeks would it take for them to run out of shoes?

14. CJ's family restaurant has a seating limit of 99 customers. If 23 groups of 4 people need to be seated at the restaurant, will there be enough seats for everyone to sit at the same time?

Use mental math to solve.

1. 40)80   40)800   40)8,000   40)80,000

2. 70)140   70)1,400   70)14,000   70)140,000

3. 10)90   10)900   10)9,000   10)90,000

4. 60)120   60)1,200   60)12,000   60)120,000

Solve. Use multiplication to check.

5. 3 1)9 8

6. 1 5)4 0

7. 4 5)3 7 5

Check

Check

Check

Solve.

8. 6 2)4 4 9

9. 7 9)6 3 3

10. 2 5)6 8

11. 5 8)2 3 9

12. 8 0)2 4 0

13. 1 4)8 8

14. 3 9)3 5 2

15. 6 9)5 5 2

16. 3)6 9 3

17. 5)1,0 5 5

18. 2)8 4 6

19. 4)8 0

20. 6)1 2 6

2-Digit Quotients

Complete the problem. Mark the answer.

1.

$$26\overline{)213}$$ with quotient 7, -182

○ Adjust the quotient up.
○ Adjust the quotient down.
○ The quotient is correct.

2.

$$34\overline{)199}$$ with quotient 6, -204

○ Adjust the quotient up.
○ Adjust the quotient down.
○ The quotient is correct.

Solve. Use multiplication to check.

3.
$$47\overline{)845}$$

Check

4.
$$22\overline{)365}$$

Check

Solve and label your answer.

5. Principal Payne purchased pencils in packs of 24. He purchased a total of 600 pencils. How many packs did he purchase?

Solve

6. Hand sanitizer is sold in 360-milliliter containers. Each 360-milliliter container holds 12 fluid ounces. How many milliliters are in a fluid ounce?

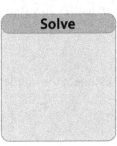
Solve

Solve.

7.
$$7\overline{)\$4.97}$$

8.
$$44\overline{)805}$$

9.
$$12\overline{)95}$$

10.
$$21\overline{)648}$$

11.
$$12\overline{)384}$$

12.
$$21\overline{)675}$$

13.
$$37\overline{)713}$$

14.
$$43\overline{)953}$$

Solve and label your answer.

1. A total of 240 cookies were made for the school open house. There were 80 people at the open house. How many cookies could each person have?

2. Miss Feldman was setting up her classroom for the new school year. She had 7 books to place on each student's desk, for a total of 175 books. How many desks were in the classroom?

3. Jackson practiced his division facts 5 days last week for a total of 150 minutes. How many minutes of practice did he average each day?

4. Grace Christian School sold candy bars to raise money for new desks. Twelve classes sold a total of 4,824 candy bars. If they each sold an equal amount, how many candy bars did each class sell?

Complete the problem. Mark the answer.

5.
$$\begin{array}{r} 3 \\ 17\overline{)69} \\ -51 \end{array}$$
○ Adjust the quotient up.
○ Adjust the quotient down.
○ The quotient is correct.

6.
$$59\overline{)175}$$ with 3 above
○ Adjust the quotient up.
○ Adjust the quotient down.
○ The quotient is correct.

7.
$$\begin{array}{r} 21 \\ 23\overline{)499} \\ -46 \\ \hline 39 \\ -23 \end{array}$$
○ Adjust the quotient up.
○ Adjust the quotient down.
○ The quotient is correct.

8.
$$\begin{array}{r} 5 \\ 41\overline{)249} \\ -205 \end{array}$$
○ Adjust the quotient up.
○ Adjust the quotient down.
○ The quotient is correct.

Round the divisor. Estimate the quotient.

9. $49\overline{)253}$

10. $63\overline{)127}$

11. $19\overline{)189}$

12. $81\overline{)569}$

Divide.

13. $61\overline{)558}$

14. $32\overline{)419}$

15. $20\overline{)620}$

16. $12\overline{)721}$

17. $48 \div 8 =$ _____

18. $40 \div 5 =$ _____

19. $56 \div 7 =$ _____

20. $54 \div 9 =$ _____

21. $80 \div 8 =$ _____

22. $90 \div 10 =$ _____

23. $32 \div 4 =$ _____

24. $30 \div 5 =$ _____

25. $24 \div 8 =$ _____

26. $63 \div 7 =$ _____

4-Digit Dividends

Name _____

Solve the division problem. Write the answers.

1.
> 405 crayons
> 24 crayons in a box
> _____ boxes of crayons
> _____ crayons left over

2.
> 5,210 staples
> 50 staples per stapler
> _____ staplers filled with staples
> _____ staples left over

Solve. Use multiplication to check.

3. $18\overline{)5{,}778}$

4. $34\overline{)3{,}089}$

5. $29\overline{)1{,}077}$

6. $56\overline{)2{,}135}$

Check

Check

Check

Check

Solve and label your answer.

7. Bradley bought a carton of milk every day for 21 days and spent $18.69 total. How much did he spend per carton of milk?

8. Mr. Coleman has 1,068 inches of trim to finish his construction project. He needs 89 feet. Does he have enough trim?

Solve. Fill in the missing digits.

9.
```
          □ □ r □ □
  4 7 ) 3, 9 □ 3
      - 3 □ 6
          □ 0 3
        - 1 □ 8
            1 5
```

10.
```
            □ 4 r □ 2
  6 7 ) 1, 0 0 0
      -   6 7
          □ 3 0
        - 2 6 □
            □ 2
```

Use the table to complete the line graph and answer the questions.

Jared's Quiz Scores	
Quiz	Score
1	90
2	70
3	85
4	95
5	100
6	90
7	90
8	95

1.

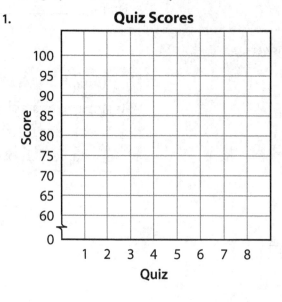

Quiz Scores

2. Which quiz has the lowest score? _____

3. Which quiz has the highest score? _____

4. The quiz scores range from _____ to _____.

5. What is the average score for quizzes 1–5? _____

Complete the table.

6.

Rule: _____	
Input	Output
18	2
27	3
45	5
81	
90	

7.

Rule: _____	
Input	Output
50	5
80	8
100	10
1,000	
10,000	

8.

Rule: _____	
Input	Output
80	2
120	3
240	6
480	
1,200	

9.

Rule: _____	
Input	Output
50	1
250	5
750	15
1,000	
1,500	

Solve.

10. $55 \div 11 =$ _____

11. $21 \div 7 =$ _____

12. $14 \div 7 =$ _____

13. $48 \div 4 =$ _____

14. $32 \div 8 =$ _____

15. $10 \overline{)200}$

16. $20 \overline{)40}$

17. $30 \overline{)120}$

18. $5 \overline{)25}$

19. $7 \overline{)42}$

20. $54 \div 9 =$ _____

21. $7 \overline{)63}$

22. $54 \div 6 =$ _____

23. $8 \overline{)56}$

24. $28 \div 7 =$ _____

3-Digit Quotients

Name _____

Complete the table. Answer the questions.

1.

Book Type	Pages	Chapters	Average Pages per Chapter
cookbook	1,100	20	
historical fiction	1,911	13	
Bible commentary	2,640	66	

2. Which book type has the most pages? _____

3. Which book type has the fewest chapters? _____

4. Which book type has the most pages per chapter? _____

Solve. Use multiplication to check.

5.
$$12\overline{)3,745}$$

Check

6.
$$90\overline{)1,003}$$

Check

Solve and label your answer.

7. Ted's Discount Tires spent $6,156 for 54 new tires. How much did Ted spend per tire?

8. Mr. Jones ordered $4,980 in prizes for field day to be used at 30 activity booths. What will be the average amount spent for prizes per booth?

Solve.

9.
$$7\overline{)1,057}$$

10.
$$25\overline{)3,649}$$

11.
$$41\overline{)7,463}$$

12.
$$34\overline{)7,854}$$

13.
$$19\overline{)1,007}$$

14.
$$43\overline{)9,159}$$

15.
$$37\overline{)7,859}$$

16.
$$51\overline{)9,488}$$

Solve the division problem. Write the remainder.

1.
1,433 chocolate chips
45 cookies to put them in
_____ chips per cookie
_____ chips left over

2.
6,393 pencils
83 holders to put them in
_____ pencils per holder
_____ pencils left over

3.
936 bracelets
24 girls to wear them
_____ bracelets per girl
_____ bracelets left over

Solve. Use multiplication to check.

4.
82)2,471

5.
31)992

6.
47)3,439

Check

Check

Check

Solve.

7.
24)2,934

8.
40)8,000

9.
34)408

10.
81)968

Complete the table.

11.
Rule: _____	
Input	Output
6	12
10	16
15	21
30	
45	

12.
Rule: _____	
Input	Output
8	1
64	8
72	9
96	
120	

13.
Rule: _____	
Input	Output
43	38
62	57
88	83
91	
100	

14.
Rule: _____	
Input	Output
20	800
40	1,600
60	2,400
80	
100	

More 3-Digit Quotients

Name _____

Solve and label your answer.

1. The longest golf course in the world is in China and measures about 8,550 yards long. It is an 18-hole golf course. How many yards per hole does this course average?

2. Mrs. Rose challenged her class to eat an apple a day for the school year, which had 180 days. Each member of her class completed the challenge. They ate 3,780 apples total. How many students are in Mrs. Rose's class?

Mark whether the quotient is a 2- or 3-digit number.

3. $24\overline{)8,616}$
   - ○ 2-digit
   - ○ 3-digit

4. $24\overline{)528}$
   - ○ 2-digit
   - ○ 3-digit

5. $16\overline{)8,080}$
   - ○ 2-digit
   - ○ 3-digit

6. $33\overline{)3,267}$
   - ○ 2-digit
   - ○ 3-digit

Circle *yes* if the quotient has a remainder and *no* if the quotient does not have a remainder.

7. $20\overline{)4,105}$

   yes   no

8. $70\overline{)770}$

   yes   no

9. $30\overline{)3,613}$

   yes   no

10. $60\overline{)5,460}$

    yes   no

Solve.

11. $16\overline{)1,374}$

12. $39\overline{)618}$

13. $53\overline{)4,951}$

14. $42\overline{)976}$

15. $16\overline{)3,927}$

16. $34\overline{)5,882}$

17. $22\overline{)9,024}$

18. $25\overline{)9,850}$

Complete the table.

1.

Rule: × 20	
Input	Output
1	
4	
12	
20	
50	

2.

Rule: ÷ 30	
Input	Output
30	
120	
360	
1,200	
2,100	

3.

Rule: − 40	
Input	Output
40	
110	
250	
490	
1,000	

4.

Rule: + 50	
Input	Output
50	
100	
540	
790	
1,010	

Classify the angle as *acute*, *obtuse*, or *right*. Write the measure of the angle.

5.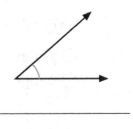

_____

_____

6.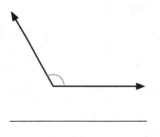

_____

_____

7.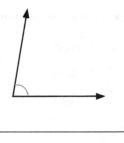

_____

_____

8.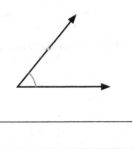

_____

_____

9.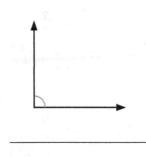

_____

_____

10.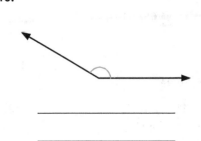

_____

_____

Solve.

11. $7\overline{)56}$

12. $8\overline{)48}$

13. $2\overline{)20}$

14. $7\overline{)21}$

15. $8\overline{)64}$

16. $63 \div 7 =$ _____

17. $48 \div 12 =$ _____

18. $55 \div 5 =$ _____

19. $30 \div 5 =$ _____

20. $72 \div 9 =$ _____

21. $9\overline{)18}$

22. $4\overline{)28}$

23. $8\overline{)40}$

24. $11\overline{)11}$

25. $12\overline{)24}$

More Division

Name _____

Solve.

1. 540 ÷ 12 = _____

2. 6,357 ÷ 28 = _____

3. 1,829 ÷ 31 = _____

4. 2,457 ÷ 91 = _____

5. 3,349 ÷ 62 = _____

6. 9,830 ÷ 42 = _____

Divide. Use multiplication to check.

7.
$$8\overline{)7,628}$$

Check

8.
$$53\overline{)5,938}$$

Check

Find the missing dividend.

9.
$$3\overline{)\,?}^{\,45}$$

_____

10.
$$17\overline{)\,?}^{\,29\ r2}$$

_____

11.
$$35\overline{)\,?}^{\,13}$$

_____

12.
$$21\overline{)\,?}^{\,69}$$

_____

13.
$$15\overline{)\,?}^{\,9\ r4}$$

_____

14.
$$9\overline{)\,?}^{\,168}$$

_____

Write the rule for the table.

15.

Rule: _____	
2	6
3	9
4	12

16.

Rule: _____	
44	11
48	12
52	13

17.

Rule: _____	
11	17
13	19
15	21

18.

Rule: _____	
11	110
50	500
100	1,000

Use the chart to answer the questions.

Verses Memorized	
Erin	
Levi	
John	
Kate	

= 10 verses

19. Who memorized the most verses? _____

20. Who memorized 30 verses? _____

21. How many more verses did Erin memorize than Kate?

_____

22. How many verses does represent?

_____

Use symbols to name the part of the circle. Answer the questions.

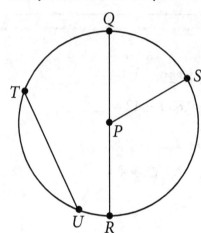

1. radius _____

2. diameter _____

3. chord _____

4. circle _____

5. How is a circle named? _____

6. What is a chord? _____

_____

7. How is a radius named? _____

_____

Write >, <, or = to compare.

8. $3\frac{3}{5}$ ◯ $3\frac{3}{8}$

9. $5\frac{4}{5}$ ◯ $5\frac{2}{10}$

10. $1\frac{1}{5}$ ◯ $1\frac{6}{25}$

11. $3\frac{3}{7}$ ◯ $3\frac{3}{8}$

12. $4\frac{12}{15}$ ◯ $4\frac{4}{5}$

13. $9\frac{6}{12}$ ◯ $9\frac{3}{8}$

14. $7\frac{6}{9}$ ◯ $7\frac{12}{18}$

15. $8\frac{6}{10}$ ◯ $8\frac{3}{5}$

16. $8\frac{3}{5}$ ◯ $8\frac{2}{7}$

17. $2\frac{7}{9}$ ◯ $2\frac{4}{8}$

18. $6\frac{2}{4}$ ◯ $6\frac{6}{12}$

19. $5\frac{1}{6}$ ◯ $5\frac{1}{4}$

Complete the division fact wheel.

20.

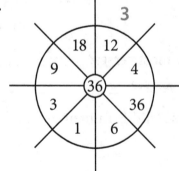

21.

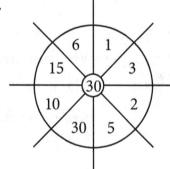

22.

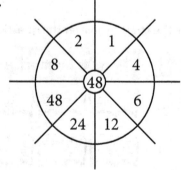

Division Review

Mark the answer.
Mark *NH* if the answer is "Not Here."

1. Kara helped her teacher cut strips of paper for a bulletin board. She cut 108 inches of paper. How many feet did she cut?

   ○ 8 ft   ○ 9 ft

   ○ 8$\frac{1}{2}$ ft   ○ NH

2. Jim is selling his miniature car collection on a website. He is selling 478 cars in sets of 5. How many complete sets can he sell?

   ○ 90   ○ 96

   ○ 94   ○ NH

3. After selling 478 cars in sets of 5, how many cars will be left over?

   ○ 2   ○ 4

   ○ 3   ○ NH

4. $$\begin{array}{r} 27\,r\square \\ 9)\overline{245} \\ -\underline{18} \\ 65 \\ -\underline{63} \\ \square \end{array}$$

   ○ 2   ○ 4

   ○ 3   ○ NH

5. $$\begin{array}{r} 65\,r\square \\ 6)\overline{391} \\ -\underline{36} \\ 31 \\ -\underline{30} \\ \square \end{array}$$

   ○ 1   ○ 3

   ○ 2   ○ NH

6. $$\begin{array}{r} ? \\ 13)\overline{\$26.00} \end{array}$$

   ○ $13.00   ○ $2.00

   ○ $20.00   ○ NH

7. Daniel baked 59 cinnamon rolls. He sold 30 to one customer and 2 packages of 6 to another. He gave 9 away to friends. How many were left over?

   ○ 1   ○ 3

   ○ 2   ○ NH

8. Emmanuel Church purchased one box of copy paper for making church bulletins. Each box contains 5,000 sheets of paper packaged into 10 packs. How many sheets are in 1 pack?

   ○ 100   ○ 50

   ○ 500   ○ NH

9. Joseph and Joshua collected their pennies throughout the year for the vacation Bible school offering. They saved 1,215 pennies and rolled them in rolls of 50. How many full rolls did they make?

   ○ 23   ○ 25

   ○ 24   ○ NH

10. How many pennies were left over after rolling up their 1,215 pennies into 50-penny rolls?

   ○ 3   ○ 15

   ○ 5   ○ NH

Solve. Mark *NH* if the answer is "Not Here."

11. $23 + (4 \times 5) =$ ___?___

   ○ 32   ○ 135
   ○ 43   ○ NH

12. $(5 + 4) \times 2 =$ ___?___

   ○ 10   ○ 32
   ○ 18   ○ NH

13. $15 + (10 \times 10) =$ ___?___

   ○ 75   ○ 275
   ○ 115   ○ NH

14. $7 + 6 - 3 =$ ___?___

   ○ 8   ○ 10
   ○ 9   ○ NH

Use mental math to solve.

15.
$20\overline{)400}$

16.
$40\overline{)16,000}$

Estimate the quotient. Solve.
Use multiplication to check.

17. **Estimate**
   $40\overline{)170}$

   Check

Round the divisor. Estimate the quotient.

18.
   $49\overline{)253}$

19.
   $71\overline{)589}$

20.
   $18\overline{)105}$

Solve.

21.
$63\overline{)378}$

22.
$33\overline{)4,093}$

23.
$18\overline{)3,912}$

24.
$56\overline{)4,931}$

Mark the number that is *not* related.

1. A. 3   C. 7
   B. 12   D. 24

2. A. $\frac{5}{12}$ C. $\frac{7}{14}$
   B. $\frac{2}{4}$ D. $\frac{10}{20}$

3. A. 16   C. 21
   B. 8   D. 56

4. A. $\frac{6}{6}$ C. $\frac{7}{7}$
   B. $\frac{11}{9}$ D. $\frac{4}{4}$

5. A. $\frac{4}{6}$ C. $\frac{1}{2}$
   B. $\frac{5}{10}$ D. $\frac{8}{16}$

6. A. 14   C. 7
   B. 49   D. 40

Mark the equivalent fraction.

7. $\frac{5}{10}$

   A. $\frac{6}{20}$ C. $\frac{3}{16}$
   B. $\frac{1}{2}$ D. $\frac{2}{15}$

8. $\frac{6}{20}$

   A. $\frac{2}{5}$ C. $\frac{1}{2}$
   B. $\frac{3}{12}$ D. $\frac{3}{10}$

9. $\frac{3}{4}$

   A. $\frac{2}{6}$ C. $\frac{6}{8}$
   B. $\frac{9}{4}$ D. $\frac{1}{2}$

10. $\frac{3}{6}$

    A. $\frac{9}{18}$ C. $\frac{10}{24}$
    B. $\frac{7}{36}$ D. $\frac{3}{21}$

11. $\frac{1}{8}$

    A. $\frac{8}{8}$ C. $\frac{8}{16}$
    B. $\frac{8}{64}$ D. $\frac{3}{21}$

12. $\frac{5}{7}$

    A. $\frac{7}{14}$ C. $\frac{15}{21}$
    B. $\frac{10}{7}$ D. $\frac{10}{40}$

Mark the number shown on the number line.

13.

   1   2   3   4   5   6   7

A. $\frac{3}{3}$ C. $4\frac{1}{2}$

B. $3\frac{1}{2}$ D. $\frac{3}{4}$

14.

   0   1

A. $\frac{3}{4}$ C. $\frac{2}{3}$

B. 3   D. $1\frac{1}{2}$

15.

        ⁻1   0   1

A. ⁻3   C. ⁻2

B. 4   D. 2

16.

   0   1

A. $\frac{1}{2}$ C. $\frac{6}{6}$

B. $\frac{2}{6}$ D. $\frac{9}{6}$

17.

 ⁻2   0   2

A. 5   C. 4

B. 2   D. ⁻4

Mark the answer.

18. $1\frac{3}{8} = \underline{\ ?\ }$

A. $\frac{24}{8}$ C. $\frac{12}{3}$

B. $\frac{11}{8}$ D. $\frac{9}{2}$

19. $\frac{11}{4} = \underline{\ ?\ }$

A. $2\frac{3}{4}$ C. $5\frac{2}{8}$

B. $15\frac{1}{4}$ D. $4\frac{1}{4}$

20. $2\frac{2}{3} = \underline{\ ?\ }$

A. $\frac{8}{3}$ C. $\frac{8}{6}$

B. $\frac{12}{14}$ D. $\frac{7}{3}$

21. $\frac{15}{6} = \underline{\ ?\ }$

A. $1\frac{1}{6}$ C. $1\frac{5}{6}$

B. $1\frac{3}{4}$ D. $2\frac{1}{2}$

22. $\frac{10}{7} = \underline{\ ?\ }$

A. $2\frac{5}{7}$ C. $1\frac{3}{7}$

B. $2\frac{1}{7}$ D. $2\frac{2}{8}$

Time

Name _____

Complete the phrase.

1.  1:09 _____ minutes after 1

2. 6:48 _____ minutes before 7

3. 2:30 half _____ two

4. 4:45 quarter to _____

Draw the time on the clock.

5.

quarter to 3

6.

nineteen minutes before 8

7.

half past 12

Fill in the blank.

8. Since 1 decade = 10 years, then 7 decades = _____ years.

9. Since 1 millennium = _____ years, then 3 millennia = 3,000 years.

10. Since 52 weeks = 1 year, then 2 years = _____ weeks.

11. Since 1 minute = _____ seconds, then 9 minutes = 540 seconds.

Write a.m. or p.m. to best match the picture.

12.

1:30 _____

13.

4:00 _____

14.

6:30 _____

Use the calendar to find the answer.

May						
S	M	T	W	Th	F	S
			1	2	3	4
5	6	7	8	9	10	11
12	13	14	15	16	17	18
19	20	21	22	23	24	25
26	27	28	29	30	31	

15. What is the date of the second Friday in May? _____

16. Which days of the week occur 5 times this month?

_____, _____, and _____

17. Mother's Day is the second Sunday in May. What is the date of Mother's Day? _____

18. What day of the week is April 30th? _____

Write the fractions from *least* to *greatest*.

1. $\dfrac{3}{4}$ $\dfrac{7}{7}$ $\dfrac{2}{4}$

_____ _____ _____

2. $\dfrac{10}{7}$ $\dfrac{1}{4}$ $\dfrac{6}{7}$

_____ _____ _____

3. $\dfrac{4}{8}$ $\dfrac{7}{5}$ $\dfrac{10}{10}$

_____ _____ _____

4. $\dfrac{1}{2}$ $\dfrac{1}{5}$ $\dfrac{1}{9}$

_____ _____ _____

5. $\dfrac{5}{3}$ $\dfrac{6}{3}$ $\dfrac{2}{3}$

_____ _____ _____

6. $\dfrac{6}{6}$ $\dfrac{12}{10}$ $\dfrac{1}{2}$

_____ _____ _____

Write >, <, or = to compare.

7. $\dfrac{4}{8} \bigcirc \dfrac{1}{2}$

8. $\dfrac{9}{9} \bigcirc \dfrac{3}{4}$

9. $\dfrac{8}{10} \bigcirc \dfrac{9}{10}$

10. $\dfrac{8}{8} \bigcirc \dfrac{1}{2}$

11. $\dfrac{10}{7} \bigcirc \dfrac{5}{10}$

12. $\dfrac{8}{12} \bigcirc \dfrac{2}{2}$

13. $\dfrac{6}{6} \bigcirc \dfrac{1}{1}$

14. $\dfrac{1}{3} \bigcirc \dfrac{3}{9}$

15. $\dfrac{3}{15} \bigcirc \dfrac{6}{1}$

16. $\dfrac{2}{3} \bigcirc \dfrac{4}{6}$

17. $\dfrac{3}{4} \bigcirc \dfrac{1}{4}$

18. $\dfrac{3}{5} \bigcirc \dfrac{2}{9}$

Solve. Use multiplication to check.

19. $48\overline{)2{,}716}$

Check

20. $83\overline{)2{,}445}$

Check

Solve.

21. $18\overline{)9{,}421}$

22. $29\overline{)7{,}399}$

23. $50\overline{)2{,}500}$

24. $63\overline{)9{,}810}$

25. $9\overline{)63}$

26. $8\overline{)56}$

27. $4\overline{)48}$

28. $4\overline{)36}$

29. $9\overline{)27}$

30. $10 \div 2 =$ ___

31. $15 \div 5 =$ ___

32. $42 \div 6 =$ ___

33. $56 \div 7 =$ ___

Elapsed Time

Name _____

Write the elapsed time.

1. Kay and her mother left for a walk at 6:45 p.m. and returned home at 7:10 p.m. _____

2. Josh and Philip went ice skating from 11:00 a.m. to 1:00 p.m. _____

3. The boys ate a snack and played a game from 1:15 p.m. to 2:25 p.m. _____

4. Maddie and Jen watched a movie from 2:10 p.m. to 4:40 p.m. _____

5. Nick slept from 10 p.m. to 8:20 a.m. _____

Add or subtract to solve. Label your answer.

Amanda cleaned for 1 hour and 45 minutes on Monday and 1 hour and 20 minutes on Saturday.

6. How long did she clean both days?

$$\begin{array}{r} 1 \text{ hr } 45 \text{ min} \\ + 1 \text{ hr } 20 \text{ min} \\ \hline \end{array}$$

_____

7. How much more time did she spend cleaning on Monday than on Saturday?

$$\begin{array}{r} 1 \text{ hr } 45 \text{ min} \\ - 1 \text{ hr } 20 \text{ min} \\ \hline \end{array}$$

_____

Rename the time.

8. 130 minutes = _____ hours _____ minutes

9. 179 minutes = _____ hours _____ minutes

Rename hours and minutes to minutes only.

10. 3 hours 14 minutes = _____ minutes

11. 6 hours 45 minutes = _____ minutes

Write the time. Label your answer.

12. Mr. Tindall umpired a baseball game that started at 6:30 p.m. The game went into overtime and lasted 3 hours and 17 minutes. What time did it end?

_____

Add or subtract. Rename if necessary.

13. $\begin{array}{r} 3 \text{ hr } 20 \text{ min} \\ + 1 \text{ hr } 50 \text{ min} \\ \hline \end{array}$

14. $\begin{array}{r} 8 \text{ hr } 25 \text{ min} \\ - 2 \text{ hr } 10 \text{ min} \\ \hline \end{array}$

15. $\begin{array}{r} 5 \text{ hr } 15 \text{ min} \\ - 3 \text{ hr } 30 \text{ min} \\ \hline \end{array}$

16. $\begin{array}{r} 1 \text{ hr } 30 \text{ min} \\ + 0 \text{ hr } 50 \text{ min} \\ \hline \end{array}$

Complete the table.

17.

Hours	Minutes
1	
4	
6	

18.

Days	Hours
1	
3	
5	

19.

Weeks	Days
2	
3	
5	

Write the factors of the number. Label the number _prime_ or _composite_.

1. 6: _____

2. 11: _____

3. 24: _____

4. 19: _____

5. 48: _____

6. 56: _____

Rename the fraction in lowest terms.

7. $\dfrac{8 \div \boxed{4}}{12 \div \boxed{4}} = \boxed{}$

8. $\dfrac{18 \div \boxed{}}{28 \div \boxed{}} = \boxed{}$

9. $\dfrac{5 \div \boxed{}}{35 \div \boxed{}} = \boxed{}$

10. $\dfrac{9 \div \boxed{}}{63 \div \boxed{}} = \boxed{}$

11. $\dfrac{6 \div \boxed{}}{8 \div \boxed{}} = \boxed{}$

12. $\dfrac{10 \div \boxed{}}{15 \div \boxed{}} = \boxed{}$

13. $\dfrac{24 \div \boxed{}}{36 \div \boxed{}} = \boxed{}$

14. $\dfrac{4 \div \boxed{}}{16 \div \boxed{}} = \boxed{}$

Solve and label your answer.

15. Karlie ate lunch at the local diner 8 times in the last three weeks. She spent a total of $68.56. What was the average amount that she spent for each meal?

16. Falls Road Baptist School used 540 reams of copy paper last school year. Their school year was 36 weeks. How many reams of paper did they average per week last school year?

17. At the end-of-the-year class party, Mrs. Todd placed 24 gumballs in each of the 19 goody bags for her class. How many gumballs did she put in the bags?

18. Mrs. Thompson has 18 students in her fifth grade class. She has graded 1,140 papers for each student this year. How many papers has she graded altogether?

Complete the equation using × or ÷.

19. $8 \bigcirc 3 = 24$

20. $4 \bigcirc 6 = 24$

21. $32 \bigcirc 8 = 4$

22. $12 \bigcirc 3 = 4$

23. $8 \bigcirc 7 = 56$

24. $49 \bigcirc 7 = 7$

25. $7 \bigcirc 2 = 14$

26. $9 \bigcirc 9 = 81$

27. $8 \bigcirc 9 = 72$

28. $63 \bigcirc 9 = 7$

Solve.

29. $\dfrac{12}{4} = $ _____

30. $6\overline{)12}$

31. $12 \cdot 5 = $ _____

32. $\dfrac{12}{12} = $ _____

33. $2\overline{)12}$

Linear Measurement

Name _____

Use a ruler to measure the length in inches. Circle the correct length.

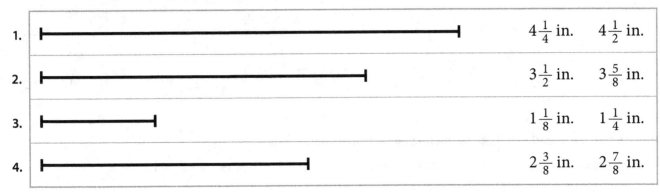

1.		$4\frac{1}{4}$ in. $4\frac{1}{2}$ in.
2.		$3\frac{1}{2}$ in. $3\frac{5}{8}$ in.
3.		$1\frac{1}{8}$ in. $1\frac{1}{4}$ in.
4.		$2\frac{3}{8}$ in. $2\frac{7}{8}$ in.

Solve and label your answer.

5. Christine and Terri helped their teacher measure a bulletin board. It was 9 feet wide and 3 feet tall. How many feet of border will they need to go around the perimeter of the board?

6. A square sticker measures 1 inch on each side. What is the perimeter of the sticker? Draw a picture and label the measurement of one side of the sticker.

Circle the better unit of measurement.

7.	the distance between the front and back of a house	inches	yards
8.	the length of your lunchbox	inches	feet
9.	the height of a fifth grader	feet	yards

Draw a picture that illustrates the indicated length.

10. a piece of rope $4\frac{1}{8}$ inches long

11. a string of beads $6\frac{3}{8}$ inches long

Complete the fact.

12. 1 ft = _____ in.

13. 1 wk = _____ days

14. 1 hr = _____ min

15. 1 mi = _____ yd

16. 1 mo = _____ days

17. 1 min = _____ sec

Rename the fraction in lowest terms.

1. $\frac{2}{4} = \frac{n}{2}$ $\frac{1}{2}$ 2. $\frac{10}{16} = \frac{n}{8}$ _____   3. $\frac{8}{36} = \frac{n}{9}$ _____   4. $\frac{18}{20} = \frac{n}{10}$ _____

5. $\frac{6}{9} = \frac{n}{3}$ _____   6. $\frac{3}{30} = \frac{n}{10}$ _____   7. $\frac{5}{15} = \frac{n}{3}$ _____   8. $\frac{10}{35} = \frac{n}{7}$ _____

9. $\frac{6}{14} = \frac{n}{7}$ _____   10. $\frac{28}{32} = \frac{n}{8}$ _____   11. $\frac{20}{24} = \frac{n}{6}$ _____   12. $\frac{4}{32} = \frac{n}{8}$ _____

Write a division equation to find the missing factor. Solve for n.

13. $12 \cdot n = 36$

$n = \underline{\;36 \div \quad\quad}$

$n = \underline{\quad\quad\quad}$

14. $6 \cdot n = 30$

$n = \underline{\quad\quad\quad}$

$n = \underline{\quad\quad\quad}$

15. $n \cdot 8 = 48$

$n = \underline{\quad\quad\quad}$

$n = \underline{\quad\quad\quad}$

16. $7 \cdot n = 56$

$n = \underline{\quad\quad\quad}$

$n = \underline{\quad\quad\quad}$

17. $n \cdot 9 = 63$

$n = \underline{\quad\quad\quad}$

$n = \underline{\quad\quad\quad}$

18. $2 \cdot n = 24$

$n = \underline{\quad\quad\quad}$

$n = \underline{\quad\quad\quad}$

Write the elapsed time or the time that is described.

19. 7:00 a.m. to 9:57 a.m.

_____

20. 10:30 a.m. to 12:10 p.m.

_____

21. 6:00 p.m. to 3:25 a.m.

_____

22. 2 hours 15 minutes after 11:00 p.m.

_____

23. 4 hours 23 minutes before 5:00 p.m.

_____

24. 6 hours 45 minutes after 8:00 a.m.

_____

Write the equivalent unit of time.

25. 1 year = _____ days

26. 1 century = _____ years

27. 1 leap year = _____ days

28. 1 week = _____ days

29. 1 hour = _____ minutes

30. 1 month = _____ days

31. 1 day = _____ hours

32. 1 year = _____ months

33. 1 minute = _____ seconds

34. 1 year = _____ weeks

35. 1 decade = _____ years

36. 1 millennium = _____ years

Renaming Measurements

Name _____

Multiply to rename larger units to smaller units.

1. 4 yd = __?__ ft

$\boxed{1 \text{ yd} = 3 \text{ ft}}$

4 × _____ = _____
4 yd = _____ ft

2. 3 mi = __?__ ft

$\boxed{1 \text{ mi} = 5{,}280 \text{ ft}}$

3 × _____ = _____
3 mi = _____ ft

3. 7 yd = __?__ in.

$\boxed{1 \text{ yd} = 36 \text{ in.}}$

7 × _____ = _____
7 yd = _____ in.

4. 3 yd = _____ ft

5. 2 mi = _____ yd

6. 4 yd = _____ in.

7. 2 mi = _____ ft

8. 5 yd = _____ ft

9. 3 mi = _____ yd

Divide to rename smaller units to larger units.

10. 60 in. = __?__ ft

$\boxed{12 \text{ in.} = 1 \text{ ft}}$

60 ÷ _____ = _____
60 in. = _____ ft

11. 144 in. = __?__ yd

$\boxed{36 \text{ in.} = 1 \text{ yd}}$

144 ÷ _____ = _____
144 in. = _____ yd

12. 6 ft = __?__ yd

$\boxed{3 \text{ ft} = 1 \text{ yd}}$

6 ÷ _____ = _____
6 ft = _____ yd

13. 9 ft = _____ yd

14. 48 in. = _____ ft

15. 108 in. = _____ yd

Divide to rename using two units.

16. 30 in. = __?__ ft __?__ in.

$\boxed{12 \text{ in.} = 1 \text{ ft}}$

30 ÷ _____ = _____
30 in. = _____ ft _____ in.

17. 7 ft = __?__ yd __?__ ft

$\boxed{3 \text{ ft} = 1 \text{ yd}}$

7 ÷ _____ = _____
7 ft = _____ yd _____ ft

18. 40 in. = _____ yd _____ in.

19. 10 ft = _____ yd _____ ft

Multiply to rename using the smaller unit.

20. 2 yd 1 ft = __?__ ft

$\boxed{1 \text{ yd} = 3 \text{ ft}}$

(2 × _____) + _____ = _____
2 yd 1 ft = _____ ft

21. 3 ft 8 in. = __?__ in.

$\boxed{1 \text{ ft} = 12 \text{ in.}}$

(3 × _____) + _____ = _____
3 ft 8 in. = _____ in.

22. 2 mi 1,000 ft = _____ ft

23. 2 yd 2 ft = _____ ft

Write the measurement for each point on the ruler.

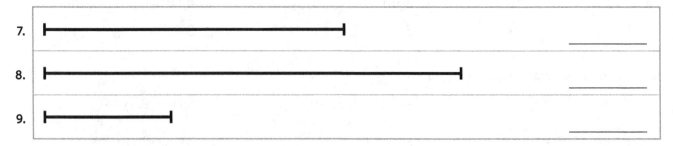

1. A = _____

2. B = _____

3. C = _____

4. D = _____

5. E = _____

6. F = _____

Use a ruler to measure the length in inches.

7. _____

8. _____

9. _____

Write the elapsed time or the time that is described.

10. | 1:00 a.m. to 6:23 a.m. |

_____

11. | 10:30 p.m. to 1:05 a.m. |

_____

12. | 3:05 p.m. to 8:37 p.m. |

_____

13. | 3 hr 20 min before 6:00 p.m. |

_____

14. | 10 hr 15 min after 10:08 p.m. |

_____

15. | 1 hr 10 min before 1:00 p.m. |

_____

Write the missing number.

16.
$$\begin{array}{r} 6 \\ \boxed{}\overline{)24} \end{array}$$

17.
$$\begin{array}{r} 3 \\ 5\overline{)\boxed{}} \end{array}$$

18.
$$\begin{array}{r} \boxed{} \\ 9\overline{)81} \end{array}$$

19.
$$\begin{array}{r} 4 \\ \boxed{}\overline{)32} \end{array}$$

20.
$$\begin{array}{r} 9 \\ 8\overline{)\boxed{}} \end{array}$$

21.
$$\begin{array}{r} 12 \\ -\ \boxed{} \\ \hline 6 \end{array}$$

22.
$$\begin{array}{r} 11 \\ \times\ \boxed{} \\ \hline 66 \end{array}$$

23.
$$\begin{array}{r} \boxed{} \\ -\ 2 \\ \hline 7 \end{array}$$

24.
$$\begin{array}{r} 8 \\ +\ 6 \\ \hline \boxed{} \end{array}$$

25.
$$\begin{array}{r} \boxed{} \\ \times\ 2 \\ \hline 24 \end{array}$$

26.
$$\begin{array}{r} 10 \\ \times\ 3 \\ \hline \boxed{} \end{array}$$

27.
$$\begin{array}{r} \boxed{} \\ +\ 7 \\ \hline 9 \end{array}$$

28.
$$\begin{array}{r} 11 \\ -\ 7 \\ \hline \boxed{} \end{array}$$

29.
$$\begin{array}{r} 7 \\ +\ \boxed{} \\ \hline 10 \end{array}$$

30.
$$\begin{array}{r} 13 \\ -\ \boxed{} \\ \hline 8 \end{array}$$

31.
$$\begin{array}{r} \boxed{} \\ -\ 8 \\ \hline 7 \end{array}$$

32.
$$\begin{array}{r} \boxed{} \\ +\ 8 \\ \hline 13 \end{array}$$

33.
$$\begin{array}{r} \boxed{} \\ \times\ 5 \\ \hline 45 \end{array}$$

34.
$$\begin{array}{r} 9 \\ +\ \boxed{} \\ \hline 15 \end{array}$$

Weight & Capacity

Name _____

Write the best unit of measurement.

1. _____ cat
2. _____ car
3. _____ hummingbird
4. _____ child
5. _____ elephant
6. _____ mouse
7. _____ man
8. _____ potatoes

oz
lb
tn

Write the best unit of measurement.

9. _____ pitcher of tea
10. _____ glass of milk
11. _____ sour cream for tacos
12. _____ jar of jelly
13. _____ water in a tub
14. _____ water for 1 serving of cocoa

c
pt
qt
gal

Rename the units.

Multiply to rename larger units to smaller units.

15. 12 gal = _?_ qt

  (1 gal = 4 qt)

  12 × _____ = _____

  12 gal = _____ qt

16. 3 qt = _?_ c

  (1 qt = 4 c)

  3 × _____ = _____

  3 qt = _____ c

17. 3 lb = _____ oz
18. 3 tn = _____ lb

Divide to rename smaller units to larger units.

19. 18 c = _?_ pt

  (2 c = 1 pt)

  18 ÷ _____ = _____

  18 c = _____ pt

20. 20 qt = _?_ gal

  (4 qt = 1 gal)

  20 ÷ _____ = _____

  20 qt = _____ gal

21. 12 c = _____ qt
22. 80 oz = _____ lb

Rename using 2 units.

23. 9 qt = _?_ gal _?_ qt

  (4 qt = 1 gal)

  9 ÷ _____ = _____

  9 qt = _____ gal _____ qt

24. 40 oz = _?_ lb _?_ oz

  (16 oz = 1 lb)

  40 ÷ _____ = _____

  40 oz = _____ lb _____ oz

25. 15 c = _____ pt _____ c
26. 10 qt = _____ gal _____ qt

Complete the fact.

27. 1 c = _____ oz
28. 1 gal = _____ qt
29. 1 qt = _____ c
30. 1 qt = _____ pt
31. 1 lb = _____ oz
32. 1 tn = _____ lb

Rename the units.

1. 12 in. = _____ ft
2. 36 in. = _____ yd
3. 5 yd = _____ ft
4. 4 yd = _____ in.
5. 1 mi = _____ ft
6. 48 in. = _____ ft
7. 180 ft = _____ yd
8. 3 yd = _____ ft
9. 1 yd = _____ ft
10. 3 mi = _____ ft
11. 72 in. = _____ yd
12. 3 yd = _____ in.

Use the information to fill in the elapsed time.

	Airline	Flight Number	Departure Time	Arrival Time	Elapsed Time
13.	Air France	9113	4:35 p.m.	6:04 p.m.	
14.	Alaska Airlines	6565	10:42 p.m.	2:33 a.m.	
15.	Delta Air Lines	1894	6:15 p.m.	3:14 a.m.	
16.	Korean Air	7094	7:54 a.m.	8:43 a.m.	
17.	Lufthansa	1494	3:34 p.m.	10:46 p.m.	

Complete the table.

18.

Days	Hours
1	
2	
3	

19.

Feet	Inches
2	
4	
6	

20.

Seconds	Minutes
120	
240	
360	

Solve.

21. $\begin{array}{r} 345 \\ \times\ \ \ 4 \\ \hline \end{array}$

22. 54 ÷ 9 = _____

23. $\frac{28}{4}$ = _____

24. 6 • 9 = _____

25. 63 ÷ 7 = _____

26. 9 • 5 = _____

27. $\begin{array}{r} 473 \\ \times\ \ \ 6 \\ \hline \end{array}$

28. $\frac{49}{7}$ = _____

Temperature

Name _____

Circle the thermometer that matches the description. Write the Fahrenheit temperature.

1. Jon checked the thermometer and put a jacket on. The temperature was _____.

2. At recess, Jon was getting warm and took his jacket off. The temperature was _____.

3. After school, Jon changed into short sleeves to ride his bike. The temperature was _____.

Write the temperature.

4. Jon's mom was boiling water to make pasta for dinner. The boiling water was _____°F.

5. To keep the freezer working properly, Jon's father set it below _____°F.

Write >, <, or = to compare.

6. 1 decade ◯ 100 yr

7. 5,280 ft ◯ 1 mi

8. 2,765 yd ◯ 1 mi

9. 10 mo ◯ 1 yr

10. 24 in. ◯ 1 yd

11. 1 tn ◯ 1,200 lb

12. 3 c ◯ 2 pt

13. 50 wk ◯ 1 yr

14. 1 leap year ◯ 365 days

Use the chart to answer the questions.

7-Day Forecast						
Tuesday	Wednesday	Thursday	Friday	Saturday	Sunday	Monday
56°F	76°F	65°F	75°F	86°F	83°F	74°F

15. Which day and temperature will be the warmest? _____

16. Which day and temperature will be the coolest? _____

17. What is the difference between the highest and lowest temperatures? _____

18. How much will the temperature decrease between Wednesday and Thursday? _____

19. How would this 7-day forecast help you make plans for the week's activities?

_____

Write the best unit of measurement.

oz	lb	tn

1. _____

2. _____

3. _____

4. _____

5. _____

6. _____

Mark the better unit of measure.

7. ◯ c ◯ gal

8.  ◯ c ◯ gal

9. ◯ c ◯ qt

10. a container of milk for a family ◯ qt ◯ gal

11. a container of milk for a child ◯ c ◯ gal

12. water in a swimming pool ◯ qt ◯ gal

Rename the units.

13. 3 lb = _____ oz

14. 3 ft = _____ in.

15. 3 hr = _____ min

16. 2 gal = _____ qt

17. 32 oz = _____ lb

18. 5 pt = _____ c

19. 3 days = _____ hr

20. 4 qt = _____ pt

21. 48 in. = _____ ft

22. 4,000 lb = _____ tn

23. 1 tn = _____ lb

24. 8 oz = _____ c

Complete the equation using +, −, ×, or ÷.

25. $8 \bigcirc 2 = 16$

26. $18 \bigcirc 9 = 9$

27. $560 \bigcirc 7 = 80$

28. $9 \bigcirc 4 = 36$

29. $8 \bigcirc 6 = 48$

30. $14 \bigcirc 8 = 6$

31. $280 \bigcirc 4 = 70$

32. $7 \bigcirc 5 = 12$

33. $63 \bigcirc 9 = 7$

34. $15 \bigcirc 3 = 18$

35. $45 \bigcirc 10 = 55$

36. $21 \bigcirc 3 = 7$

Measurement Problems

Name _____

Use the chart to solve. Rename if necessary.

1. What is the total weight of Bob's fish?

   $$\begin{array}{r} 4\text{ lb }2\text{ oz} \\ +\ 7\text{ lb }7\text{ oz} \\ \hline \end{array}$$

2. What is the total weight of Tim's fish?

Bob's Catch	Tim's Catch
19 in. 4 lb 2 oz	18 in. 3 lb 6 oz
23 in. 7 lb 7 oz	21 in. 5 lb 11 oz

3. How much more does Bob's catch weigh than Tim's?

4. Did either fisherman catch a fish longer than 2 feet? If so, who?

Solve. Rename if necessary. Label your answer.

5. Calvin and Curt mailed packages to their dad in the military. Each boy sent a package that weighed 2 pounds 5 ounces. What was the total weight of the boxes?

6. When driving to the lake, Mr. Hughes travels at an average speed of 60 miles per hour. What distance does he travel in 5 hours?

7. Brian purchased 2 pallets of sod for his yard. Each pallet weighs 1,800 pounds. The crane can hold 4 tons. Will it be able to lift both pallets at once? Why or why not?

8. Mrs. Fisher purchased a 64-ounce bottle of laundry detergent. After a week of laundry, she had 48 ounces left. How many cups did she use?

Solve. Rename if necessary.

9. $\begin{array}{r} 6\text{ ft }\ 3\text{ in.} \\ +\ 1\text{ ft }10\text{ in.} \\ \hline \end{array}$

10. $\begin{array}{r} 3\text{ hr }40\text{ min} \\ -\ 1\text{ hr }50\text{ min} \\ \hline \end{array}$

11. $\begin{array}{r} 5\text{ lb }\ 8\text{ oz} \\ +\ 2\text{ lb }12\text{ oz} \\ \hline \end{array}$

12. $\begin{array}{r} 6\text{ gal }1\text{ qt} \\ -\ 2\text{ gal }3\text{ qt} \\ \hline \end{array}$

Write the Fahrenheit temperature.

1.   2.   3.   4. 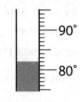 5.

_____   _____   _____   _____   _____

Shade the thermometer to show the temperature.

6. 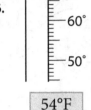 7.  8. 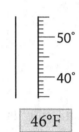 9.  10.

54°F   8°F   46°F   0°F   ⁻22°F

Solve. Rename if necessary.

11. 6 ft 3 in.
   + 5 ft 8 in.

12. 3 yd 5 in.
   − 1 yd 2 in.

13. 5 lb 9 oz
   ×   4

14. 2 gal 1 qt
   + 5 gal 2 qt

15. 7 yd 11 in.
   −   20 in.

16. 3 tn 800 lb
   ×   3

17. 8 yd 2 ft
   + 1 yd 2 ft

18. 7 hr 30 min
   − 2 hr 45 min

Rename the units.

19.

Rename as smaller units.	
2 tn	_____ lb
5 gal	_____ qt
4 yd	_____ ft
4 hr	_____ min
3 days	_____ hr
2 c	_____ oz

20.

Rename as larger units.	
6 pt	_____ qt
32 oz	_____ lb
108 in.	_____ yd
14 days	_____ wk
180 min	_____ hr
12 qt	_____ gal

21.

Rename using 2 units.		
40 oz	_____ lb	_____ oz
7 c	_____ pt	_____ c
53 hr	_____ days	_____ hr
43 in.	_____ ft	_____ in.
5 qt	_____ gal	_____ qt
2,500 lb	_____ tn	_____ lb

Complete the fact.

22. 1 week = _____ days

23. 1 ton = _____ pounds

24. 1 gallon = _____ quarts

25. 1 yard = _____ feet

26. 1 foot = _____ inches

27. 1 day = _____ hours

28. 1 year = _____ days

29. 1 mile = _____ yards

30. 1 pound = _____ ounces

© 2021 BJU Press. Reproduction prohibited.

Time & Customary Measurement Review

Name _____

Mark the answer.
Mark *NH* if the answer is "Not Here."

1. 3 days = __?__ hr
   - ○ 36
   - ○ 24
   - ○ 48
   - ○ NH

2. 16 qt = __?__ gal
   - ○ 3
   - ○ 4
   - ○ 5
   - ○ NH

3. 1 mi ○ 5,281 ft
   - ○ >
   - ○ <
   - ○ =
   - ○ NH

4. A 2-pound box of salt weighs __?__ oz.
   - ○ 12
   - ○ 16
   - ○ 32
   - ○ NH

5. The line segment below measures __?__ in.

   ├────────────┤
   - ○ 1
   - ○ $1\frac{1}{2}$
   - ○ $1\frac{1}{4}$
   - ○ NH

6. Kelly took a 75-minute nap.

   __?__ hr __?__ min
   - ○ 1 hr 5 min
   - ○ 1 hr 15 min
   - ○ 2 hr
   - ○ NH

7. Jan's 2 children each have a 45-minute piano lesson. Sarah's starts at 3 p.m. with Kimberly's following. At what time will Kimberly's lesson end?
   - ○ 3:45 p.m.
   - ○ 4:15 p.m.
   - ○ 4:30 p.m.
   - ○ NH

8. Mrs. Terrill ate lunch with her students 3 hours and 25 minutes before the school pep rally at 3:00 p.m. At what time did they eat?
   - ○ 11:35 a.m.
   - ○ 12:00 p.m.
   - ○ 12:25 p.m.
   - ○ NH

9. quarter to 5
   - ○ 3:45
   - ○ 4:15
   - ○ 5:00
   - ○ NH

10. 6 min before 11
    - ○ 10:54
    - ○ 11:06
    - ○ 11:46
    - ○ NH

11. Jared is 130 months old. On his next birthday he will be __?__ yrs old.
    - ○ 10
    - ○ 11
    - ○ 12
    - ○ NH

12. Mrs. Long started her job at 7:25 a.m. She left to go home at 3:51 p.m. How long did she work?
    - ○ 3 hr 34 min
    - ○ 4 hr 25 min
    - ○ 8 hr 26 min
    - ○ NH

13. Mr. Brondyke ran his first marathon in 4 hours 26 minutes. His tenth marathon was 2 hours 51 minutes. What is the difference?
    - ○ 1 hr 35 min
    - ○ 2 hr 25 min
    - ○ 3 hr 30 min
    - ○ NH

14. Grandpa Nutz just had a birthday. He is 9 decades old. How old is he?
    - ○ 80
    - ○ 90
    - ○ 95
    - ○ NH

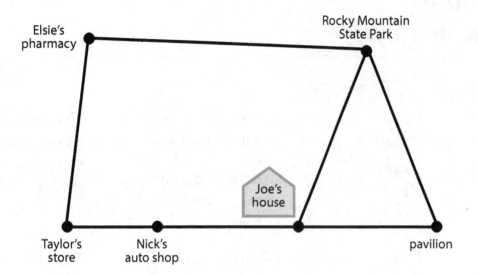

Elsie's pharmacy

Rocky Mountain State Park

Joe's house

Taylor's store Nick's auto shop pavilion

Key

1 in. = 3 mi

Use the key to complete the table.

15.

Inches	1	2	3	4	5
Miles					

Use a ruler, the map, and the key to find the answers.

16. The distance from Taylor's store to Elsie's pharmacy is _____ miles.

17. It is _____ miles from Rocky Mountain State Park to Joe's house.

18. What is the distance from the pavilion to Nick's auto shop? _____ miles

19. What is the distance around the perimeter of the map? _____ miles

Complete the table.

20.

qt	pt
2	
4	
6	
8	

21.

ft	yd
3	
9	
12	
27	

22.

tn	lb
1	
2	
3	
5	

23.

hr	days
24	
72	
120	
240	

Write the Fahrenheit temperature.

24.

—40°

—30°

_____

25.

—0°

—-10°

_____

Circle the answer.

26.

normal body temperature

98.6°F   32°F   212°F

27.

boiling point of water

98.6°F   32°F   212°F

Mark the answer.

1.
$$\begin{array}{r} 57 \\ \times\ 9 \\ \hline \end{array}$$
A. 503
B. 513
C. 613
D. 623

2.
$$\begin{array}{r} 345 \\ \times\ \ 8 \\ \hline \end{array}$$
A. 276
B. 1,276
C. 2,470
D. 2,760

3.
$$\begin{array}{r} \$4.15 \\ \times\ \ \ \ \ 6 \\ \hline \end{array}$$
A. $24.90
B. $249.00
C. $24.30
D. $2.49

4.
$$\begin{array}{r} 51 \\ \times 24 \\ \hline \end{array}$$
A. 1,300
B. 1,002
C. 1,224
D. 2,040

5. $3 \times 27 =$
A. 270
B. 3 + 20 + 7
C. 300 + 27
D. 3 × (20 + 7)

6. $4 \times 3,000 =$
A. 120
B. 1,200
C. 12,000
D. 120,000

7. $5,863 \times 0 =$
A. 1
B. 5,863
C. 0
D. 5,000

8. $54 \div 9 =$
A. 5
B. 6
C. 7
D. 8

9. $8\overline{)64}$
A. 5
B. 6
C. 7
D. 8

10. $5\overline{)315}$
A. 63
B. 73
C. 630
D. 6,300

11. $6\overline{)3,000}$
A. 5
B. 50
C. 500
D. 5,000

12. $6,300 \div 7 =$
A. 9
B. 90
C. 900
D. 9,000

13. $\frac{28}{4} =$
A. 5
B. 7
C. 9
D. 11

Use plane *d* to identify the figure.

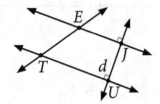

14. $\overleftrightarrow{TE}$

    A. point
    B. line
    C. angle
    D. plane

15. $\angle JET$

    A. point
    B. line
    C. angle
    D. plane

16. $\angle JUT$

    A. acute angle
    B. obtuse angle
    C. right angle
    D. straight angle

17. $\angle ETU$

    A. acute angle
    B. obtuse angle
    C. right angle
    D. straight angle

18. $\angle ETU$

    A. $\overrightarrow{TE}$ and $\overrightarrow{TU}$
    B. $\overrightarrow{UT}$ and $\overrightarrow{UJ}$
    C. $\overrightarrow{ET}$ and $\overrightarrow{EJ}$
    D. $\overleftrightarrow{TU}$ and $\overleftrightarrow{EJ}$

Mark the answer.

19.

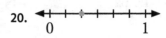

$\dfrac{3}{4} =$

    A. $\dfrac{1}{4}$
    B. $\dfrac{2}{4}$
    C. $\dfrac{6}{8}$
    D. $\dfrac{8}{8}$

20.

0   1

    A. $\dfrac{1}{5}$
    B. $\dfrac{2}{6}$
    C. $\dfrac{1}{2}$
    D. 2

21. $\dfrac{2}{7} = \dfrac{n}{14}$

    A. 4
    B. $1\dfrac{3}{4}$
    C. 7
    D. 14

22. $\dfrac{21}{21} =$

    A. 1
    B. 21
    C. 2
    D. 12

23. $\dfrac{8}{10} =$

    A. $\dfrac{2}{10}$
    B. $\dfrac{1}{10}$
    C. $\dfrac{4}{5}$
    D. $\dfrac{10}{10}$

24. $1\dfrac{2}{5} =$

    A. $\dfrac{3}{5}$
    B. $\dfrac{5}{5}$
    C. $\dfrac{8}{5}$
    D. $\dfrac{7}{5}$

Adding Like Fractions

Name _____

Solve.

1. $\frac{2}{5} + \frac{2}{5} =$

2. $\frac{3}{9} + \frac{2}{9} =$

3. $\frac{1}{8} + \frac{2}{8} =$

4. $\frac{3}{7} + \frac{3}{7} =$

Add. Write the answer in lowest terms.

5. $\begin{array}{r} \frac{1}{4} \\ + \frac{1}{4} \\ \hline \frac{2 \div \boxed{2}}{4 \div \boxed{2}} = \end{array}$

6. $\begin{array}{r} \frac{1}{6} \\ + \frac{2}{6} \\ \hline \frac{\div \boxed{3}}{\div \boxed{3}} = \end{array}$

7. $\begin{array}{r} \frac{4}{8} \\ + \frac{2}{8} \\ \hline \end{array}$

8. $\begin{array}{r} \frac{5}{12} \\ + \frac{4}{12} \\ \hline \end{array}$

9. $\begin{array}{r} \frac{1}{12} \\ + \frac{1}{12} \\ \hline \end{array}$

10. $\begin{array}{r} \frac{3}{6} \\ + \frac{1}{6} \\ \hline \end{array}$

11. $\begin{array}{r} \frac{2}{9} \\ + \frac{1}{9} \\ \hline \end{array}$

12. $\begin{array}{r} \frac{2}{12} \\ + \frac{1}{12} \\ \hline \end{array}$

13. $\begin{array}{r} \frac{2}{10} \\ + \frac{2}{10} \\ \hline \end{array}$

14. $\begin{array}{r} \frac{2}{9} \\ + \frac{4}{9} \\ \hline \end{array}$

Estimate. Add. Write the answer in lowest terms.

15. **Estimate** $\begin{array}{r} 2\frac{1}{6} \\ + 1\frac{3}{6} \\ \hline \end{array}$

16. **Estimate** $\begin{array}{r} 3\frac{8}{10} \\ + 2\frac{4}{10} \\ \hline \end{array}$

17. **Estimate** $\begin{array}{r} 5\frac{2}{5} \\ + 2\frac{3}{5} \\ \hline \end{array}$

Solve. Write the answer in lowest terms.

18. $\begin{array}{r} \frac{4}{5} \\ + \frac{3}{5} \\ \hline \end{array}$

19. $\begin{array}{r} \frac{5}{10} \\ + \frac{2}{10} \\ \hline \end{array}$

20. $\begin{array}{r} \frac{4}{5} \\ + \frac{1}{5} \\ \hline \end{array}$

21. $\begin{array}{r} \frac{4}{8} \\ + \frac{2}{8} \\ \hline \end{array}$

22. $\begin{array}{r} \frac{3}{10} \\ + \frac{3}{10} \\ \hline \end{array}$

23. $\frac{1}{4} + \frac{3}{4} + \frac{3}{4} =$

24. $\frac{1}{6} + \frac{3}{6} + \frac{1}{6} =$

25. $\frac{1}{7} + \frac{1}{7} + \frac{5}{7} =$

26. $\frac{1}{2} + \frac{1}{2} + \frac{1}{2} =$

Use the table to find the answer. Solve. Show your work.

1. What will it cost to rent a car for the month of May at the daily rate?

   _____

Car Rental Rates	
daily	$45.50
weekly	$169.90
monthly	$595.90

2. What will it cost to rent a car for 8 weeks at the weekly rate?

   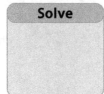

   _____

3. What will it cost to rent a car for 2 months at the monthly rate?

   _____

4. If you wanted to rent a car for 8 weeks (2 months), would the weekly rate or the monthly rate be less expensive?

   _____

Plot and label the points on the graph.

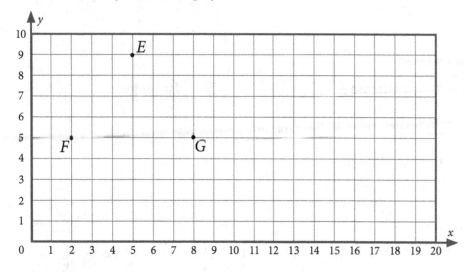

5. A (10, 2)

6. B (18, 2)

7. C (15, 7)

8. D (12, 9)

9. Draw $\overleftrightarrow{AC}$.

10. Draw $\overrightarrow{BD}$.

Write the ordered pair for the points. Draw the figure.

11. E _____

12. F _____

13. G _____

14. Draw $\overline{EG}$.

15. Draw $\overline{GF}$.

16. Draw $\overline{FE}$.

Write >, <, or = to compare.

17. $\frac{1}{2}$ ◯ $\frac{7}{8}$

18. $\frac{1}{2}$ ◯ $\frac{3}{6}$

19. $\frac{2}{4}$ ◯ $\frac{3}{10}$

20. $\frac{6}{6}$ ◯ $\frac{1}{2}$

21. $\frac{3}{6}$ ◯ $\frac{5}{10}$

22. $\frac{7}{7}$ ◯ $\frac{6}{9}$

23. $\frac{4}{5}$ ◯ $\frac{1}{9}$

24. $\frac{3}{4}$ ◯ $\frac{2}{10}$

Subtracting Like Fractions

Name _____

Subtract.

1. $\dfrac{2}{3}$
    $-\dfrac{1}{3}$
    ___

2. $\dfrac{5}{8}$
    $-\dfrac{2}{8}$
    ___

3. $\dfrac{4}{10}$
    $-\dfrac{1}{10}$
    ___

4. $\dfrac{7}{9}$
    $-\dfrac{3}{9}$
    ___

5. $\dfrac{5}{6}$
    $-\dfrac{4}{6}$
    ___

Subtract. Write the answer in lowest terms.

6. $\dfrac{4}{6}$
    $-\dfrac{2}{6}$
    ___

    $\dfrac{2 \div \boxed{2}}{6 \div \boxed{2}} =$

7. $\dfrac{7}{12}$
    $-\dfrac{4}{12}$
    ___
    $\dfrac{\div \boxed{3}}{\div \boxed{3}} =$

8. $\dfrac{8}{10}$
    $-\dfrac{4}{10}$
    ___

9. $\dfrac{8}{9}$
    $-\dfrac{4}{9}$
    ___

10. $\dfrac{2}{4}$
    $-\dfrac{1}{4}$
    ___

11. $\dfrac{4}{8} - \dfrac{3}{8} =$

12. $\dfrac{7}{9} - \dfrac{3}{9} =$

13. $\dfrac{3}{4} - \dfrac{1}{4} =$

14. $\dfrac{4}{6} - \dfrac{1}{6} =$

Subtract. Write the answer in lowest terms. Rename if necessary.

15. $^{1}2\dfrac{1}{3}^{4}$
    $-\ 1\dfrac{2}{3}$
    ___

16. $1\dfrac{2}{5}$
    $-\ \dfrac{4}{5}$
    ___

17. $2\dfrac{3}{9}$
    $-\ \dfrac{6}{9}$
    ___

18. $\dfrac{4}{6}$
    $-\ \dfrac{1}{6}$
    ___

Estimate. Solve. Write the answer in lowest terms.

19. **Estimate**
    $5\dfrac{8}{9}$
    $-\ 3\dfrac{2}{9}$
    ___

20. **Estimate**
    $3\dfrac{5}{8}$
    $-\ 1\dfrac{4}{8}$
    ___

21. **Estimate**
    6
    $-\ 3\dfrac{4}{12}$
    ___

Solve. Write the answer in lowest terms.

22. $3\dfrac{3}{5}$
    $+\ 1\dfrac{2}{5}$
    ___

23. $2\dfrac{5}{7}$
    $+\ 3\dfrac{4}{7}$
    ___

24. $4\dfrac{4}{10}$
    $-\ 2\dfrac{5}{10}$
    ___

25. $2\dfrac{1}{4}$
    $-\ 1\dfrac{3}{4}$
    ___

Use the pictograph to answer the question.

Dave's Cycle Shop took a color inventory of all their motorcycles. They listed the results in this pictograph.

Motorcycle Colors	
red	
white	
blue	
green	
yellow	
black	
other	

= 2 motorcycles

1. How many yellow motorcycles are on the lot?

_____

2. How many more white motorcycles are there than green ones?

_____

3. Explain the symbol 🏍.

_____

_____

4. Explain what the word "other" means in the last row of colors.

_____

_____

5. Dave sold all the red and "other" color motorcycles. How many did he sell?

_____

Multiply or divide by a name for 1 to find an equivalent fraction.

6. $\dfrac{3 \times \boxed{2}}{9 \times \boxed{2}} = \dfrac{}{18}$

7. $\dfrac{6 \div \boxed{2}}{8 \div \boxed{2}} = \dfrac{}{4}$

8. $\dfrac{2}{3} = \dfrac{}{15}$

9. $\dfrac{12}{16} = \dfrac{}{8}$

10. $\dfrac{1}{2} = \dfrac{}{6}$

11. $\dfrac{4}{8} = \dfrac{}{24}$

12. $\dfrac{20}{30} = \dfrac{}{3}$

13. $\dfrac{14}{28} = \dfrac{}{4}$

14. $\dfrac{7}{8} = \dfrac{}{16}$

15. $\dfrac{4}{10} = \dfrac{}{15}$

16. $\dfrac{6}{16} = \dfrac{}{8}$

17. $\dfrac{1}{3} = \dfrac{}{24}$

18. $\dfrac{9}{36} = \dfrac{}{4}$

19. $\dfrac{5}{25} = \dfrac{}{5}$

20. $\dfrac{3}{10} = \dfrac{}{20}$

Write the numbers from *least* to *greatest*.

21. | ⁻5 | 4 | 3 | ⁻2 |

____ ____ ____ ____

22. | ⁻2 | ⁻1 | 17 | ⁻33 |

____ ____ ____ ____

23. | ⁻4 | 4 | ⁻2 | ⁻6 |

____ ____ ____ ____

24. | ⁻8 | 4 | 33 | ⁻20 |

____ ____ ____ ____

25. | ⁻17 | ⁻10 | 46 | ⁻20 |

____ ____ ____ ____

26. | 11 | 8 | ⁻10 | ⁻3 |

____ ____ ____ ____

Adding Unlike Fractions

Name _____

Add. Rename to make like fractions.

1. $\dfrac{3}{8} =$

   $+ \dfrac{2 \times \boxed{2}}{4 \times \boxed{2}} =$

2. $\dfrac{1}{6}$

   $+ \dfrac{2}{3}$ **1**

3. $\dfrac{1}{4}$

   $+ \dfrac{1}{2}$

4. $\dfrac{2}{4}$

   $+ \dfrac{1}{12}$

Add. Rename if needed. Write the answer in lowest terms.

5. $\dfrac{2}{10} =$

   $+ \dfrac{2}{5} =$

   $\dfrac{6 \div \boxed{2}}{10 \div \boxed{2}} =$

6. $\dfrac{3}{12}$

   $+ \dfrac{2}{6}$

7. $\dfrac{1}{4}$

   $+ \dfrac{2}{8}$

8. $\dfrac{3}{9}$

   $+ \dfrac{2}{9}$

Solve. Rename if needed. Write the answer in lowest terms.

9. $\dfrac{3}{6}$

   $+ \dfrac{1}{3}$

10. $\dfrac{3}{10}$

    $+ \dfrac{3}{5}$

11. $\dfrac{3}{8}$

    $+ \dfrac{2}{4}$

12. $\dfrac{2}{6}$

    $+ \dfrac{1}{12}$

13. $2\dfrac{1}{3}$

    $+ 4\dfrac{3}{6}$

14. $3\dfrac{1}{4}$

    $+ 1\dfrac{1}{2}$

15. $\dfrac{3}{5}$

    $+ \dfrac{7}{10}$

16. $6\dfrac{5}{6}$

    $+ 2\dfrac{2}{3}$

Subtract. Rename if needed. Write the answer in lowest terms.

17. $\dfrac{4}{8}$

    $- \dfrac{3}{8}$

18. 2

    $- 1\dfrac{3}{5}$

19. $5\dfrac{1}{7}$

    $- 3\dfrac{2}{7}$

20. $\dfrac{4}{12} - \dfrac{2}{12} =$

21. $\dfrac{6}{10} - \dfrac{1}{10} =$

Mark all the equivalent fractions.

22. ○ $\dfrac{1}{3}$
    ○ $\dfrac{2}{5}$
    ○ $\dfrac{2}{6}$

23. ○ $\dfrac{2}{5}$
    ○ $\dfrac{4}{10}$
    ○ $\dfrac{6}{8}$

24. ○ $\dfrac{4}{12}$
    ○ $\dfrac{1}{3}$
    ○ $\dfrac{2}{6}$

25. ○ $\dfrac{3}{6}$
    ○ $\dfrac{1}{3}$
    ○ $\dfrac{3}{9}$

Add or subtract. Rename if needed. Write the answer in lowest terms.

1. $\frac{3}{8}$
   $+ \frac{1}{8}$

2. $3\frac{2}{3}$
   $+ 1\frac{1}{3}$

3. $4\frac{1}{10}$
   $+ 2\frac{2}{5}$

4. $\frac{6}{12}$
   $+ \frac{5}{6}$

5. 2
   $+ 1\frac{3}{8}$

6. $\frac{9}{12}$
   $- \frac{6}{12}$

7. $2\frac{1}{8}$
   $- 1\frac{2}{8}$

8. 4
   $- 1\frac{9}{12}$

9. $\frac{7}{11}$
   $- \frac{3}{11}$

10. $\frac{4}{6}$
    $- \frac{2}{6}$

Write an improper fraction and a mixed number for the picture.

11.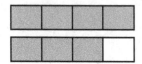

_____ _____
improper   mixed
fraction   number

12.

_____ _____
improper   mixed
fraction   number

13. $\frac{2}{3} + \frac{2}{3} + \frac{2}{3} + \frac{2}{3}$

_____ _____
improper   mixed
fraction   number

Draw a picture for the phrase.

14. $\frac{9}{3}$ of the pizzas

15. $\frac{5}{2}$ of the pizzas

Match.

_____ 16. 493,205,750

_____ 17. 943,520,507

_____ 18. 394,251,175

_____ 19. 123,456,789

_____ 20. 987,852,963

A number with the greatest value

B The value of 4 is 400,000.

C 493 millions, 205 thousands, 750 ones

D nine hundred forty-three million, five hundred twenty thousand, five hundred seven

E 300,000,000 + 90,000,000 + 4,000,000 + 200,000 + 50,000 + 1,000 + 100 + 70 + 5

Subtracting Unlike Fractions

Name _____

Subtract. Write the answer in lowest terms. Mark a fractional piece of the figure with an *X* to show subtraction.

1. $\dfrac{3}{4} =$
$-\dfrac{1 \times 2}{2 \times 2} =$

2. $\dfrac{1}{2}$
$-\dfrac{2}{6}$

3. $\dfrac{7}{8}$
$-\dfrac{3}{4}$

Find the missing number to make an equivalent fraction.

4. $\dfrac{1}{2} = \dfrac{}{6}$

5. $\dfrac{3}{4} = \dfrac{}{12}$

6. $\dfrac{2}{5} = \dfrac{}{10}$

7. $\dfrac{5}{6} = \dfrac{}{18}$

8. $\dfrac{2}{8} = \dfrac{}{24}$

9. $\dfrac{4}{10} = \dfrac{}{20}$

10. $\dfrac{1}{3} = \dfrac{}{9}$

11. $\dfrac{1}{7} = \dfrac{}{21}$

Solve. Rename if needed. Write the answer in lowest terms.

12. $\dfrac{3 \times 2}{4 \times 2}$
$-\dfrac{1}{8}$

13. $\dfrac{4}{6}$
$-\dfrac{1}{3}$

14. $\dfrac{7}{9}$
$-\dfrac{2}{3}$

15. $\dfrac{1}{5}$
$-\dfrac{1}{10}$

16. $3\dfrac{5}{6}$
$-1\dfrac{2}{6}$

17. $5\dfrac{1}{2}$
$-4\dfrac{1}{4}$

18. $4\dfrac{2}{3}$
$-1\dfrac{2}{9}$

19. $2\dfrac{6}{8}$
$-1\dfrac{1}{4}$

Add or subtract. Write the answer in lowest terms.

20. $\dfrac{1}{3} + \dfrac{1}{3} + \dfrac{1}{3} =$

21. $\dfrac{1}{5} + \dfrac{2}{5} + \dfrac{4}{5} =$

22. $\dfrac{1}{9} + \dfrac{1}{9} + \dfrac{1}{9} =$

23. $\dfrac{10}{12}$
$-\dfrac{3}{6}$

24. $2\dfrac{1}{9}$
$-1\dfrac{1}{9}$

25. 4
$+2\dfrac{2}{3}$

26. 5
$-3\dfrac{3}{5}$

Write the equation. Solve and label.

1. Sue's granola bar recipe calls for $2\frac{1}{4}$ cups of almonds and $3\frac{1}{2}$ cups of walnuts. How many cups of nuts are needed for her recipe?

2. The boys ate $3\frac{1}{3}$ pans of brownies. The girls ate $1\frac{5}{6}$ pans of brownies. How many more pans of brownies did the boys eat than the girls?

Write >, <, or = to compare.

3. $4\frac{2}{4} \bigcirc 4\frac{6}{9}$

4. $8\frac{5}{15} \bigcirc 7\frac{2}{14}$

5. $1\frac{3}{6} \bigcirc 1\frac{7}{14}$

6. $\frac{10}{35} \bigcirc \frac{18}{20}$

7. $3\frac{1}{6} \bigcirc 3\frac{2}{8}$

8. $2\frac{4}{8} \bigcirc 2\frac{10}{16}$

9. $1\frac{8}{10} \bigcirc 1\frac{2}{12}$

10. $3\frac{3}{5} \bigcirc 2\frac{2}{7}$

Write the mixed numbers from *least* to *greatest*.

11. $2\frac{8}{16}$ $1\frac{1}{2}$ $1\frac{4}{12}$

12. $5\frac{1}{8}$ $5\frac{6}{8}$ $5\frac{9}{16}$

13. $4\frac{8}{10}$ $11\frac{3}{10}$ $4\frac{2}{5}$

____ ____ ____

____ ____ ____

____ ____ ____

Complete the table.

14.

Rule: + 7	
Input	Output
5	
14	
33	
41	
58	
60	
77	

15.

Rule: – 8	
Input	Output
10	
23	
38	
45	
59	
71	
100	

16.

Rule: + 10	
Input	Output
55	
99	
105	
143	
291	
310	
505	

17.

Rule: – 5	
Input	Output
143	
295	
301	
444	
515	
831	
1,000	

Least Common Multiple

Follow the steps to add the fractions.

$$\frac{2 \times 2}{6 \times 2} = \frac{4}{12}$$
$$+ \frac{3 \times 3}{4 \times 3} = \frac{9}{12}$$
$$\frac{13}{12} = 1\frac{1}{2}$$

1. List the first five nonzero multiples of each denominator.

   6: _____

   4: _____

2. Circle the least common multiple of 6 and 4.

Mark all correct answers.

3. How was the least common multiple used to solve the above problem?

   ○ The least common multiple (12) became the common denominator.

   ○ The fractions were renamed by multiplying each by the least common multiple (12).

4. Why was $\frac{2}{6}$ multiplied by $\frac{2}{2}$ and $\frac{3}{4}$ by $\frac{3}{3}$?

   ○ Fractions were renamed by multiplying by a fractional name for 1.

   ○ This gave equivalent fractions with denominators of 12 (LCM).

List nonzero multiples of the numbers to find the least common multiple.
Solve. Write the answer in lowest terms.

5. $\frac{2 \times 3}{5 \times 3} =$
   $+ \frac{2 \times 5}{3 \times 5} =$

   3: _____

   5: _____

6. $\frac{2}{9}$
   $+ \frac{1}{6}$

7. $\frac{3}{4}$
   $+ \frac{4}{5}$

8. $\frac{3}{4}$
   $- \frac{1}{10}$

   4: _____

   10: _____

9. $\frac{1}{2}$
   $- \frac{3}{7}$

10. $\frac{6}{9}$
    $- \frac{1}{12}$

Complete the table. Write the answers in lowest terms.

11.

Rule: $+ \frac{1}{2}$	
Input	Output
$\frac{1}{2}$	
$\frac{1}{4}$	
1	

12.

Rule: $+ 2\frac{1}{3}$	
Input	Output
$\frac{1}{3}$	
2	
$\frac{2}{3}$	

13.

Rule: $+ 3\frac{1}{4}$	
Input	Output
$1\frac{3}{4}$	
$\frac{1}{2}$	
$2\frac{1}{4}$	

Solve. Write the answer in lowest terms.

1. $\frac{3}{4}$
   $+ \frac{1}{4}$

2. $\frac{7}{5}$
   $- \frac{3}{5}$

3. $2\frac{1}{3}$
   $+ 1\frac{5}{6}$

4. $4\frac{2}{4}$
   $- 2\frac{3}{8}$

Estimate. Solve. Write the answer in lowest terms.

5. Estimate
   $4\frac{6}{8}$
   $+ 2\frac{3}{8}$

6. Estimate
   $3\frac{5}{10}$
   $+ 1\frac{2}{10}$

7. Estimate
   $7\frac{1}{3}$
   $+ 6\frac{2}{3}$

Use mental math to solve. Label your answer.

8. Mr. Winstead's toolbox contains 1,600 screws. He has divided them evenly into 40 piles. How many screws are in each pile?

9. Mrs. Nolan used 8 slices of bread to make 4 sandwiches (2 slices per sandwich). She cut the sandwiches into 4 pieces. How many sandwich pieces did she prepare?

10. The choir at Falls Road School took 800 orders for their annual Valentine balloon sale. There are 20 students in the choir. How many orders did each student take if each took the same number?

11. The local radio station received 4,000 toys for their annual Christmas toy drive. The toys will be sent to needy children in 20 countries. How many toys will be sent to each country if the same number is sent to each?

Write the first 5 nonzero multiples of the number.

12. 3: _____

13. 7: _____

14. 5: _____

15. 8: _____

16. 6: _____

17. 10: _____

Divide.

18. $30\overline{)60}$

19. $40\overline{)160}$

20. $50\overline{)2,500}$

21. $80\overline{)40,000}$

22. $30\overline{)1,200}$

23. $9\overline{)72}$

24. $8\overline{)56}$

25. $7\overline{)63}$

26. $8\overline{)32}$

27. $2\overline{)16}$

Comparing Fractions

Name _____

Write >, <, or = to compare.

1. $\frac{3}{4} \bigcirc \frac{6}{8}$

2. $\frac{4}{5} \bigcirc \frac{5}{7}$

3. $\frac{1}{3} \bigcirc \frac{2}{4}$

4. $\frac{3}{4} \bigcirc \frac{8}{12}$

5. $\frac{1}{2} \bigcirc \frac{3}{5}$

6. $\frac{5}{6} \bigcirc \frac{3}{4}$

7. $\frac{4}{9} \bigcirc \frac{1}{3}$

8. $\frac{4}{10} \bigcirc \frac{3}{5}$

Use the information to answer the questions.

The Jackson children took part in a horse show. There were 12 jumps in the show-jumping category. Jon completed $\frac{5}{12}$ of the jumps, Melanie completed $\frac{5}{6}$ of the jumps, and Abby completed $\frac{3}{4}$ of the jumps.

9. Write a math statement to compare Jon's and Melanie's jumps. _____

10. Which of the Jacksons completed the most jumps? _____

Mark the answer.

11. Which statement correctly compares Melanie's and Abby's jumps?

$\bigcirc \frac{5}{6} < \frac{3}{4}$ $\bigcirc \frac{5}{6} = \frac{3}{4}$ $\bigcirc \frac{5}{6} > \frac{3}{4}$

12. Which statement correctly compares Jon's and Abby's jumps?

$\bigcirc \frac{5}{12} < \frac{3}{4}$ $\bigcirc \frac{5}{12} = \frac{3}{4}$ $\bigcirc \frac{5}{12} > \frac{3}{4}$

Write an equivalent fraction to represent the number of jumps completed out of 12.

If each person attempted 12 jumps, how many jumps did each complete?

13. Abby: $\frac{3}{4} = \frac{}{12}$ _____ jumps

14. Jon: $\frac{5}{12} = $ _____ jumps

15. Melanie: $\frac{3}{6} = $ _____ jumps

Solve. Write the answer in lowest terms.

16. $\frac{1}{3} + \frac{2}{3} + \frac{2}{3} =$

17. $\frac{1}{4} + \frac{3}{4} + \frac{2}{4} =$

18. $\frac{3}{5} + \frac{1}{5} + \frac{1}{5} =$

19. $\frac{5}{6} + \frac{5}{6} + \frac{5}{6} =$

© 2021 BJU Press. Reproduction prohibited.

Math 5 Activities

Solve. Rename if needed. Write the answer in lowest terms.

1. $\frac{3}{7}$
    $+ \frac{2}{7}$

2. $\frac{7}{8}$
    $+ \frac{7}{8}$

3. $\frac{2}{4}$
    $+ \frac{3}{8}$

4. $\frac{3}{6}$
    $- \frac{1}{3}$

5. $\frac{5}{9}$
    $- \frac{1}{3}$

6. $\frac{2}{3}$
    $+ \frac{1}{5}$

7. $\frac{7}{9}$
    $- \frac{3}{4}$

8. 4
    $- 3\frac{4}{7}$

9. $3\frac{3}{10}$
    $- \frac{7}{10}$

Solve. Use multiplication to check.

10. $23\overline{)8,743}$

    Check

11. $15\overline{)177}$

    Check

12. $88\overline{)6,169}$

    Check

Solve.

13. $8\overline{)\$25.04}$

14. $81\overline{)6,492}$

15. $16\overline{)1,183}$

Complete the equation to find the measure of the unknown angle.

16.

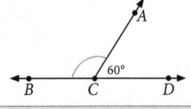

$\angle ACD = 60°$ $\angle ACB = \underline{\ ?\ }$

_____ + n = 180°

n = 180° − _____

n = _____

$\angle ABC$ = _____

17.

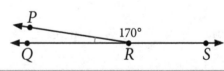

$\angle PRS = 170°$ $\angle PRQ = \underline{\ ?\ }$

_____ + n = 180°

n = 180° − _____

n = _____

$\angle PRQ$ = _____

Least Common Denominator

List the multiples until the least common multiple (LCM) is found. Circle the LCM.

1. 3: _3, 6, 9, 12, (15)_
   5: _5, 10, (15)_

2. 6: _____
   9: _____

3. 8: _____
   10: _____

Solve. Write the answer in lowest terms.

4. $\frac{1}{4} = \frac{3}{12}$
   $+ \frac{4}{6} = \frac{8}{12}$
   _____

5. $\frac{1}{5} = \frac{}{15}$
   $+ \frac{2}{3} = \frac{}{15}$
   _____

6. $\frac{3}{4}$
   $- \frac{1}{3}$
   _____

7. $\frac{1}{2}$
   $+ \frac{1}{5}$
   _____

8. $\frac{1}{2}$
   $+ \frac{2}{6}$
   _____

9. $\frac{5}{9}$
   $- \frac{1}{3}$
   _____

10. $\frac{9}{14}$
    $- \frac{2}{7}$
    _____

11. $\frac{3}{4}$
    $- \frac{3}{6}$
    _____

12. $\frac{2}{3}$
    $+ \frac{1}{4}$
    _____

13. $\frac{7}{20}$
    $- \frac{1}{5}$
    _____

14. $\frac{5}{21}$
    $+ \frac{3}{7}$
    _____

15. $\frac{7}{18}$
    $- \frac{2}{9}$
    _____

16. $\frac{4}{5}$
    $+ \frac{2}{3}$
    _____

17. $\frac{8}{9}$
    $- \frac{1}{6}$
    _____

18. $\frac{5}{8}$
    $- \frac{2}{10}$
    _____

19. $\frac{1}{15}$
    $+ \frac{3}{5}$
    _____

20. $\frac{1}{2} + \frac{1}{3} + \frac{1}{6} =$

21. $\frac{1}{10} + \frac{1}{5} + \frac{1}{2} =$

22. $\frac{1}{2} + \frac{1}{4} + \frac{1}{8} =$

Complete the table.

23.

Rule: $+ \frac{1}{3}$	
Input	Output
$\frac{2}{3}$	
1	
$\frac{1}{3}$	

24.

Rule: $- \frac{1}{4}$	
Input	Output
1	
$\frac{3}{4}$	
$\frac{1}{4}$	

25.

Rule: $+ \frac{3}{5}$	
Input	Output
$\frac{1}{5}$	
$\frac{4}{5}$	
$\frac{2}{5}$	

Solve and label your answer.

> The Kelly family spent their vacation in the mountains. They hiked a mountain trail and fished in the river.

1. Alex hiked $\frac{3}{4}$ of the trail. Macey hiked $\frac{1}{2}$ of the trail. How much more of the trail did Alex hike than Macey?

   _____

2. Dad hiked the whole trail $2\frac{1}{4}$ times. How much farther did Dad hike than Alex and Macey combined?

   _____

3. Dad caught $5\frac{1}{2}$ pounds of fish, Alex caught $2\frac{3}{4}$ pounds, and Macey caught $3\frac{5}{8}$ pounds. How many pounds of fish did they catch altogether?

   _____

Solve and label your answer.

4. Brian and Ann went to visit their family in North Carolina for Memorial Day. They left Greenville, SC, at 9:15 a.m. They arrived in Rocky Mount, NC, at 3:20 p.m. How long did they travel?

5. Lori went to the local grocery store to purchase some snack food to munch on in the car on her trip home. She got in the checkout line at 2:20 p.m. and finished paying at 2:32 p.m. How much time did she spend in line?

Write the time that is described.

6.
| 3 hr 20 min |
| after 7:45 a.m. |

   _____

7.
| 1 hr 57 min |
| before 12:30 p.m. |

   _____

8.
| 7 hr 10 min |
| after 10:00 a.m. |

   _____

Write a division equation to find the missing factor. Solve for *n*.

9. $8 \times n = 56$

   $n = \underline{\;56 \div\;}$

   $n = \underline{\qquad}$

10. $6 \times n = 48$

    $n = \underline{\qquad}$

    $n = \underline{\qquad}$

11. $n \times 3 = 36$

    $n = \underline{\qquad}$

    $n = \underline{\qquad}$

12. $n \times 8 = 64$

    $n = \underline{\qquad}$

    $n = \underline{\qquad}$

13. $9 \times n = 45$

    $n = \underline{\qquad}$

    $n = \underline{\qquad}$

14. $n \times 3 = 18$

    $n = \underline{\qquad}$

    $n = \underline{\qquad}$

Adding & Subtracting Unlike Fractions

Name _____

Add or subtract. Write the answer in lowest terms.

1. $\begin{array}{r} \frac{1}{2} \\ + \frac{1}{7} \\ \hline \end{array}$

2. $\begin{array}{r} \frac{3}{4} \\ - \frac{4}{12} \\ \hline \end{array}$

3. $\begin{array}{r} \frac{15}{16} \\ - \frac{5}{8} \\ \hline \end{array}$

4. $\begin{array}{r} \frac{1}{4} \\ + \frac{5}{6} \\ \hline \end{array}$

5. $\begin{array}{r} \frac{4}{5} \\ - \frac{1}{3} \\ \hline \end{array}$

6. $\begin{array}{r} \frac{11}{21} \\ - \frac{1}{7} \\ \hline \end{array}$

7. $\begin{array}{r} \frac{5}{7} \\ - \frac{1}{4} \\ \hline \end{array}$

8. $\begin{array}{r} \frac{14}{15} \\ + \frac{3}{5} \\ \hline \end{array}$

Solve and label your answer.

9. Mrs. White needs to make a double batch of brownies for the class party. If each batch makes sixteen brownies, will she have enough brownies for 30 students?

10. Write the amounts Mrs. White will need for each ingredient as she doubles the recipe.

Brownies		
	Single Batch	**Double Batch**
flour	$1\frac{1}{2}$ c	
sugar	$\frac{3}{4}$ c	
cocoa	$\frac{1}{2}$ c	
eggs	2	

11. On Saturday Joel rode his bike $4\frac{3}{8}$ miles, and Josh rode $3\frac{1}{2}$ miles. How many miles did they ride altogether?

12. Nathan rollerbladed $1\frac{3}{4}$ miles to his friend's house. Elisabeth rode her bike $2\frac{1}{2}$ miles to school. How much farther did Elisabeth travel than Nathan?

Rewrite the fractions, using the least common denominator. Write >, <, or = to compare.

13. $\frac{4}{9}$ $\frac{2}{3}$

$\frac{4}{9}$ $<$ $\frac{6}{9}$

14. $\frac{7}{10}$ $\frac{2}{5}$

$\bigcirc$

15. $\frac{8}{15}$ $\frac{3}{5}$

$\bigcirc$

16. $\frac{1}{4}$ $\frac{3}{20}$

$\bigcirc$

Add. Write the answer in lowest terms.

1. $\frac{7}{20} + \frac{8}{20} + \frac{5}{20} =$

2. $\frac{1}{6} + \frac{3}{6} + \frac{5}{6} =$

3. $\frac{7}{18} + \frac{5}{18} + \frac{10}{18} =$

Use the Venn diagram to answer the question.

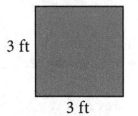

3   4

3   6
9   15   12   4   8
        24
18 21   16 20

4. The common multiples of 3 and 4 are _____ and _____.

5. Why is 20 not a common multiple of 3 and 4?

_____

Find the perimeter of the figure.

6.

3 ft

3 ft

perimeter = _____

7.

5 in.

3 in.

4 in.

perimeter = _____

8.

8 ft

3 ft

perimeter = _____

9.

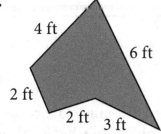

4 ft

6 ft

2 ft

2 ft   3 ft

perimeter = _____

10.

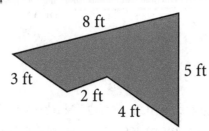

8 ft

3 ft

2 ft

4 ft

5 ft

perimeter = _____

11.

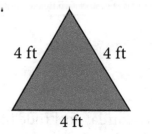

4 ft   4 ft

4 ft

perimeter = _____

Circle the fractions that are equal to 1.

12. $\frac{4}{8}$ $\frac{6}{6}$ $\frac{5}{10}$ $\frac{4}{12}$ $\frac{8}{8}$

$\frac{10}{10}$ $\frac{1}{1}$ $\frac{3}{3}$ $\frac{1}{3}$ $\frac{2}{20}$

Circle the fractions that are equal to $\frac{1}{2}$.

13. $\frac{10}{20}$ $\frac{6}{12}$ $\frac{2}{14}$ $\frac{3}{4}$ $\frac{1}{5}$

$\frac{4}{8}$ $\frac{1}{1}$ $\frac{2}{4}$ $\frac{1}{2}$ $\frac{3}{6}$

Math 5 Activities

Adding & Subtracting Mixed Numbers

Name _____

Add. Write the answer in lowest terms.

1. $4\frac{5}{6}$
$+ \ 2\frac{1}{6}$

2. $3\frac{2}{3}$
$+ \ 1\frac{2}{3}$

3. $5\frac{2}{5}$
$+ \ 3\frac{1}{5}$

4. $2\frac{3}{10}$
$+ \ 6\frac{1}{5}$

5. $3\frac{5}{12}$
$+ \ 1\frac{2}{4}$

6. $2\frac{1}{7}$
$+ \ 3\frac{5}{14}$

Subtract. Rename if needed.

7. $3\frac{3}{6}$
$- \ 1\frac{1}{6}$

8. $5\frac{7}{9}$
$- \ 3\frac{2}{3}$

9. $2\frac{7}{8}$
$- \ 1\frac{3}{16}$

10. $2\frac{9}{20}$
$- \ 1\frac{4}{5}$

11. $5\frac{15}{16}$
$- \ 2\frac{7}{8}$

12. $4\frac{1}{9}$
$- \ 2\frac{2}{3}$

Add or subtract. Rename if needed. Write the answer in lowest terms.

13. $2\frac{3}{5}$
$+ \ 1\frac{2}{3}$

14. $9\frac{2}{3}$
$- \ 1\frac{1}{4}$

15. $10\frac{1}{8}$
$- \ 2\frac{3}{4}$

16. $8\frac{3}{12}$
$- \ 6\frac{3}{4}$

17. 3
$+ \ 2\frac{5}{6}$

18. 5
$- \ 2\frac{5}{7}$

Complete the equation.

19. $3 - \boxed{} = 2\frac{1}{4}$

20. $\frac{3}{10} + \boxed{} = 1$

21. $\frac{5}{9} + \boxed{} = \frac{8}{9}$

22. $\boxed{} - \frac{5}{15} = 0$

23. $\frac{1}{2} + \frac{4}{6} = \boxed{} + \frac{1}{2}$

24. $(\frac{4}{6} + \frac{1}{3}) + \frac{2}{3} = \boxed{} + (\frac{1}{3} + \frac{2}{3})$

Solve. Write the answer in lowest terms.

1. $\frac{3}{5} + \frac{4}{5} =$

2. $\frac{1}{3} + \frac{2}{3} + \frac{2}{3} =$

Solve. Cross out fractional parts to show subtraction.

3.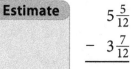

$\frac{5}{8} - \frac{4}{8} =$

4.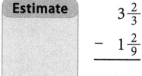

$\frac{4}{5} - \frac{1}{5} =$

5.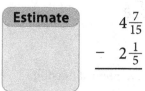

$\frac{4}{6} - \frac{3}{6} =$

Estimate. Solve. Write the answer in lowest terms.

6. | Estimate |

$5\frac{5}{12}$
$-\ 3\frac{7}{12}$
_____

7. | Estimate |

$3\frac{2}{3}$
$-\ 1\frac{2}{9}$
_____

8. | Estimate |

$4\frac{7}{15}$
$-\ 2\frac{1}{5}$
_____

Rename the units of length.

9. 24 in. = _____ ft

10. 5,280 ft = _____ mi

11. 3 ft = _____ yd

12. 72 in. = _____ yd

13. 36 in. = _____ ft

14. 1 mi = _____ ft

15. 3 yd = _____ ft

16. 120 ft = _____ yd

17. 2 yd = _____ in.

18. 3 yd = _____ in.

19. 9 ft = _____ yd

20. 108 in. = _____ yd

Complete the number sentence.

21. 38 in. = _____ ft _____ in.

22. 4 ft 8 in. = _____ in.

23. 82 in. = _____ yd _____ in.

24. 8 yd 2 ft = _____ ft

25. 8 ft = _____ yd _____ ft

26. 1 mi 5 yd = _____ yd

Write the factors of the number. Label the number *prime* or *composite*.

27. 2: _____

_____

28. 8: _____

_____

29. 14: _____

_____

30. 15: _____

_____

31. 29: _____

_____

32. 56: _____

_____

Adding & Subtracting Fractions

Name _____

Solve. Write the answer in lowest terms.

1. $\begin{array}{r} \frac{2}{3} \\ + \ \frac{4}{5} \\ \hline \end{array}$

2. $\begin{array}{r} 3\frac{2}{4} \\ - \ 1\frac{3}{8} \\ \hline \end{array}$

3. $\begin{array}{r} 7\frac{6}{7} \\ + \ 2\frac{3}{14} \\ \hline \end{array}$

4. $\begin{array}{r} \frac{15}{18} \\ - \ \frac{3}{6} \\ \hline \end{array}$

5. $\frac{1}{12} + \frac{5}{6} =$

6. $\frac{3}{18} - \frac{1}{9} =$

7. $\frac{7}{20} - \frac{3}{10} =$

Use the chart to answer the questions.

8. Which student ran the most laps on the track?

_____

9. How many more laps did Maddie run than Andrew?

_____

10. How many more laps did the girls run than the boys?

_____

11. How many laps did the students run altogether? _____

Running Practice	
Student	Laps
Andrew	$6\frac{1}{2}$
Maddie	$10\frac{3}{4}$
Allen	5
Jackie	$9\frac{1}{2}$

Write the rule for the table.

12.

Rule: _____

Input	Output
$\frac{1}{2}$	1
$\frac{1}{4}$	$\frac{3}{4}$
$\frac{3}{4}$	$1\frac{1}{4}$

13.

Rule: _____

Input	Output
2	$1\frac{1}{2}$
$\frac{3}{4}$	$\frac{1}{4}$
$\frac{1}{2}$	0

14.

Rule: _____

Input	Output
$2\frac{1}{4}$	$2\frac{1}{2}$
$1\frac{1}{8}$	$1\frac{3}{8}$
$\frac{3}{4}$	1

Add or subtract. Write the answer in lowest terms.

15. $\begin{array}{r} 3\frac{1}{2} \\ - \ 2\frac{1}{12} \\ \hline \end{array}$

16. $\begin{array}{r} 5\frac{4}{5} \\ + \ 2\frac{2}{10} \\ \hline \end{array}$

17. $\begin{array}{r} 3\frac{6}{7} \\ - \ 1\frac{5}{21} \\ \hline \end{array}$

18. $\begin{array}{r} 8\frac{7}{15} \\ - \ 1\frac{2}{3} \\ \hline \end{array}$

19. $\begin{array}{r} \frac{1}{3} = \frac{}{15} \\ \frac{1}{5} = \frac{}{15} \\ + \ \frac{1}{15} = \frac{}{15} \\ \hline \end{array}$

20. $\begin{array}{r} \frac{1}{4} \\ \frac{1}{8} \\ + \ \frac{1}{16} \\ \hline \end{array}$

21. $\begin{array}{r} \frac{2}{6} \\ \frac{7}{12} \\ + \ \frac{2}{24} \\ \hline \end{array}$

22. $\begin{array}{r} \frac{3}{7} \\ \frac{5}{14} \\ + \ \frac{9}{28} \\ \hline \end{array}$

Repartition the top figure to show the renaming. Solve.

1.

$\frac{1}{3}$	$\frac{1}{3}$	$\frac{1}{3}$

$\frac{1}{6}$	$\frac{1}{6}$	$\frac{1}{6}$	$\frac{1}{6}$	$\frac{1}{6}$	$\frac{1}{6}$

$$\begin{array}{r} \frac{2}{3} \\ + \frac{5}{6} \\ \hline \end{array}$$

2.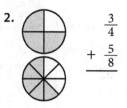

$$\begin{array}{r} \frac{3}{4} \\ + \frac{5}{8} \\ \hline \end{array}$$

Circle the equivalent fractions.

3. $\frac{4}{8}$ $\frac{3}{4}$ $\frac{5}{10}$ $\frac{42}{84}$

4. $\frac{4}{15}$ $\frac{4}{5}$ $\frac{12}{15}$ $\frac{2}{5}$

5. $\frac{3}{7}$ $\frac{2}{5}$ $\frac{4}{9}$ $\frac{6}{14}$

Write the first 5 nonzero multiples of the numbers.

6. 6: _____

    7: _____

    8: _____

    9: _____

Rename the units.

7. 32 oz = _____ lb

8. 2,000 lb = _____ tn

9. 3 lb = _____ oz

10. 16 qt = _____ gal

11. 3 pt = _____ c

12. 2 c = _____ fl oz

13. 2 hr = _____ min

14. 180 min = _____ hr

15. 360 min = _____ hr

16. 12 in. = _____ ft

17. 36 in. = _____ ft

18. 5 ft = _____ in.

Order of Operations		
1. parentheses	2. multiplication or division	3. addition or subtraction

Write the sequence for solving the equation. Solve.

19. $39 - 4 \times 6 + 8 =$ __23__
  __multiply, subtract,__
        __add__

20. $19 - 21 \div 3 =$ _____

21. $35 - 6 \times (1 + 2) =$ _____

22. $18 \times 2 \div 12 =$ _____

23. $4 + 12 \div 3 - 5 =$ _____

24. $(60 + 3) \div 9 =$ _____

Solve. Remember to use the order of operations.

25. $10 - 12 \div 3 =$ _____

26. $5 \times 3 \div 3 =$ _____

27. $5 + 6 \times 3 \div 2 =$ _____

28. $(5 + 10) \times 5 =$ _____

29. $63 \div 7 \div 3 =$ _____

30. $36 - 3 \times 4 =$ _____

31. $5 \times (3 + 1) =$ _____

32. $42 \div (3 + 4) =$ _____

33. $6 + 10 \times 1 =$ _____

34. $21 \div 7 \times 3 =$ _____

35. $8 - 4 \times 2 =$ _____

36. $6 + 4 - 3 =$ _____

Factoring to Compare Fractions

Name _____

> 1. List the prime factors from least to greatest for each denominator. Circle pairs of factors that are common to both numbers.
> 2. Multiply the common factors to find the greatest common factor (GCF).
> 3. Multiply the greatest common factor (GCF) by all uncommon factors to find the least common multiple (LCM).
> 4. Use the least common multiple (LCM) to find the equivalent fractions. Compare.

Follow the steps to compare the fractions.

1. $\frac{8}{21}$ ◯ $\frac{13}{28}$

```
        21                  26
       / \                 / \
      3 × 7              2 × 14
                          / \
                       2 × 2 × 7
```

21: __3__, __7__

28: __2__, __2__, __7__

GCF: __7__

LCM: $\underset{\text{GCF}}{\underline{\quad 7 \quad}}$ × $\underset{\text{uncommon factors}}{\underline{\quad 2 \quad} \times \underline{\quad 2 \quad} \times \underline{\quad 3 \quad}}$ = __84__

$\frac{8}{21}$ ×  =

$\frac{13}{28}$ ×  =

◯

2. $\frac{9}{12}$ ◯ $\frac{7}{18}$

```
        12                  18
       / \                 / \
```

12: ____, ____, ____

18: ____, ____, ____

GCF: ____ × ____ = ____

LCM: $\underset{\text{GCF}}{\underline{\quad\quad}}$ × $\underset{\text{uncommon factors}}{\underline{\quad\quad} \times \underline{\quad\quad} \times \underline{\quad\quad}}$ = ____

$\frac{9}{12}$ × =

$\frac{7}{18}$ × =

◯

Add or subtract. Write the answer in lowest terms.

3. $1\frac{3}{4}$
     $+\ 2\frac{3}{8}$
     ———

4. $3\frac{1}{3}$
     $-\ 1\frac{1}{5}$
     ———

5. $\frac{3}{10}$
     $+\ \frac{5}{14}$
     ———

6. $\frac{5}{8}$
     $+\ \frac{10}{24}$
     ———

Use the information to answer the question. Write the answer in lowest terms. Label your answer.

> Several chocolate bars were left over from a party. Each chocolate bar has 16 sections. Katlyn ate 14 sections, Christie ate 20 sections, and Macey ate 9 sections.

1. What part of a chocolate bar did Katlyn eat? _____

2. What part did Macey eat? _____

3. How many bars did Christie eat? _____

4. How many bars did they eat altogether? _____

Write > or < to compare.

5. $\frac{3}{4}$ ◯ $\frac{7}{12}$

6. $\frac{3}{5}$ ◯ $\frac{10}{15}$

7. $\frac{3}{9}$ ◯ $\frac{4}{7}$

8. $\frac{4}{8}$ ◯ $\frac{2}{3}$

Solve. Rename if needed.

9.
```
   3 ft  6 in.
+ 14 ft  8 in.
```

10.
```
   4 hr 50 min
+ 3 hr 40 min
```

11.
```
  16 lb 14 oz
-  7 lb 12 oz
```

12.
```
  7 tn 243 lb
+ 1 tn  50 lb
```

13.
```
  5 yd 2 ft
- 3 yd 1 ft
```

14.
```
  7 gal 1 qt
- 3 gal 2 qt
```

15.
```
  6 pt 4 c
- 1 pt 6 c
```

16.
```
  3 yd 2 ft
- 1 yd 4 ft
```

Solve. Use multiplication to check.

17. 81)95

Check

18. 93)659

Check

19. 62)5,432

Check

20. 22)1,431

Check

21. 18)109

Check

22. 49)31,290

Check

Fraction Review

Name _____

Mark the correct answer that is in lowest terms.

1. $\frac{3}{7} + \frac{3}{7} = \frac{?}{\_\_}$

   ○ $\frac{9}{7}$ ○ $\frac{6}{14}$ ○ $\frac{6}{7}$

2. $\frac{3}{12} + \frac{4}{12} + \frac{5}{12} = \frac{?}{\_\_}$

   ○ $\frac{12}{12}$ ○ $\frac{17}{12}$ ○ 1

3. $\frac{4}{9} + \frac{2}{9} = \frac{?}{\_\_}$

   ○ $\frac{2}{3}$ ○ $\frac{1}{2}$ ○ $\frac{6}{9}$

4. $\frac{1}{8} + \frac{5}{8} + \frac{4}{8} = \frac{?}{\_\_}$

   ○ $\frac{10}{8}$ ○ $1\frac{2}{8}$ ○ $1\frac{1}{4}$

5. $\frac{3}{8} + \frac{5}{8} = \frac{?}{\_\_}$

   ○ $\frac{8}{8}$ ○ $\frac{3}{4}$ ○ 1

6. $\frac{4}{6} - \frac{3}{6} = \frac{?}{\_\_}$

   ○ $\frac{1}{6}$ ○ $\frac{7}{6}$ ○ $\frac{7}{12}$

7. $\frac{3}{10} + \frac{2}{10} = \frac{?}{\_\_}$

   ○ $\frac{6}{10}$ ○ $\frac{1}{2}$ ○ $\frac{5}{20}$

8. $\frac{11}{15} - \frac{1}{15} = \frac{?}{\_\_}$

   ○ $\frac{12}{15}$ ○ $\frac{10}{15}$ ○ $\frac{2}{3}$

9. $\frac{3}{4} + \frac{3}{4} = \frac{?}{\_\_}$

   ○ $\frac{6}{4}$ ○ $\frac{9}{4}$ ○ $1\frac{1}{2}$

10. $\frac{8}{9} - \frac{5}{9} = \frac{?}{\_\_}$

    ○ $\frac{3}{9}$ ○ $\frac{1}{3}$ ○ $\frac{13}{9}$

Solve. Write the answer in lowest terms.

11. $4\frac{4}{8}$
    $-\ 3\frac{2}{8}$
    _____

12. $5\frac{7}{12}$
    $-\ 1\frac{1}{12}$
    _____

13. $3\frac{1}{2}$
    $-\ \ \frac{1}{2}$
    _____

14. $6\frac{4}{5}$
    $-\ 3$
    _____

Use the information to answer the question. Mark the answer.

> David ran $4\frac{3}{4}$ laps on the track, Nicholas ran $6\frac{1}{2}$ laps, and Andrew ran $5\frac{3}{8}$ laps.

15. How much farther did Andrew run than David?

    ○ $\frac{5}{8}$ laps   ○ $2\frac{1}{4}$ laps   ○ $1\frac{1}{2}$ laps

16. How much farther did Nicholas run than Andrew?

    ○ $1\frac{2}{3}$ laps   ○ $1\frac{1}{8}$ laps   ○ $1\frac{4}{8}$ laps

17. Which two boys ran the farthest?

    ○ David and Nicholas
    ○ David and Andrew
    ○ Andrew and Nicholas

18. How far did the boys run altogether?

    ○ $15\frac{1}{8}$ laps   ○ $16\frac{5}{8}$ laps   ○ $15\frac{1}{4}$ laps

Mark the number that completes the equivalent fraction.

19. $\frac{1}{2} = \frac{?}{10}$
- ○ 5
- ○ 4
- ○ 7

20. $\frac{1}{4} = \frac{?}{16}$
- ○ 2
- ○ 3
- ○ 4

21. $\frac{3}{5} = \frac{?}{15}$
- ○ 8
- ○ 9
- ○ 10

22. $\frac{5}{5} = \frac{?}{18}$
- ○ 11
- ○ 15
- ○ 24

Solve. Mark the correct answer.

23.
$$\begin{array}{r} \frac{1}{8} \\ + \frac{3}{4} \\ \hline \end{array}$$
- ○ $\frac{5}{8}$
- ○ $\frac{6}{8}$
- ○ $\frac{7}{8}$

24.
$$\begin{array}{r} \frac{4}{6} \\ + \frac{1}{2} \\ \hline \end{array}$$
- ○ $\frac{1}{6}$
- ○ $1\frac{1}{6}$
- ○ $\frac{5}{6}$

25.
$$\begin{array}{r} \frac{3}{14} \\ + \frac{2}{7} \\ \hline \end{array}$$
- ○ $\frac{5}{7}$
- ○ $\frac{6}{14}$
- ○ $\frac{1}{2}$

26.
$$\begin{array}{r} 4\frac{5}{16} \\ - 2\frac{1}{4} \\ \hline \end{array}$$
- ○ $6\frac{1}{16}$
- ○ $2\frac{1}{16}$
- ○ $2\frac{1}{4}$

Solve. Mark the correct answer.

27. Mom used 4 cups of grapes, $3\frac{1}{2}$ cups of melon, and $5\frac{3}{4}$ cups of pineapple for a fruit salad. How many cups of fruit did she use?
- ○ $13\frac{1}{4}$ c
- ○ $12\frac{1}{4}$ c
- ○ $12\frac{1}{2}$ c

28. Jack picked $3\frac{3}{4}$ baskets of strawberries. Will picked $2\frac{1}{2}$ baskets. How much more did Jack pick than Will?
- ○ $5\frac{1}{4}$
- ○ $1\frac{1}{4}$
- ○ $1\frac{1}{2}$

Mark the correct estimate.

29.
$$\begin{array}{r} 2\frac{3}{4} \\ + 2\frac{3}{8} \\ \hline \end{array}$$
- ○ 5
- ○ 4
- ○ 6

30.
$$\begin{array}{r} 3\frac{4}{12} \\ + 2\frac{7}{12} \\ \hline \end{array}$$
- ○ 5
- ○ 6
- ○ 7

31.
$$\begin{array}{r} 8\frac{3}{5} \\ - 4\frac{3}{10} \\ \hline \end{array}$$
- ○ 3
- ○ 4
- ○ 5

Mark the correct answer.

32. the sum of three-fourths and one-fourth
- ○ two-fourths
- ○ three-fourths
- ○ one

33. the difference between three-fifths and two-fifths
- ○ one-fifth
- ○ two-fifths
- ○ three-fifths

34. two-sixths more than one-sixth
- ○ one-half
- ○ one-sixth
- ○ two-sixths

35. one-seventh less than one
- ○ six-sevenths
- ○ five-sevenths
- ○ four-sevenths

Use the graph to find the answer.

Student Transportation

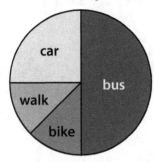

The graph shows how the 100 students at Calvary Christian School normally travel to and from school.

1. How do $\frac{1}{2}$ of the students travel to school?

   A. car

   B. bus

   C. bike

2. How do $\frac{1}{4}$ of the students travel to school?

   A. car

   B. bus

   C. bike

3. How many students travel by bus?

   A. 15

   B. 25

   C. 50

4. What type of graph is used to show the data?

   A. circle graph

   B. line graph

   C. bar graph

Average Temperatures in Seattle	
January	41°F
February	44°F
March	47°F
April	51°F

5. Which graph correctly shows the data?

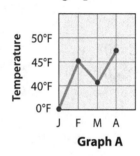
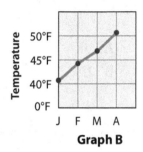

   A. line graph A

   B. line graph B

   C. neither

Lifespan of Pets

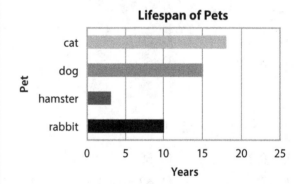

6. Which pet lives less than 5 years?

   A. cat   C. dog

   B. hamster   D. rabbit

7. About how many more years does a dog live than a rabbit?

   A. 5 years   C. 10 years

   B. 15 years   D. 20 years

Mark the answer that makes the equation true.

8. $7 \times 6 = \underline{\ ?\ } \times 7$

A. 0   C. 3

B. 1   D. 6

9. $4 + 4 + 4 = \underline{\ ?\ } \times 4$

A. 0   C. 3

B. 1   D. 5

10. $15 + \underline{\ ?\ } = 35$

A. 10   C. 30

B. 20   D. 40

11. $125 - \underline{\ ?\ } = 99$

A. 11   C. 26

B. 16   D. 30

12. $21 \times 56 = 21 \times (50 + \underline{\ ?\ })$

A. 21   C. 50

B. 56   D. 6

13. $\$1.75 - \underline{\ ?\ } = \$1.50$

A. $0.25   C. 50

B. 25   D. $2.50

14. $43 \times 6 = \underline{\ ?\ }$

A. 228   C. 258

B. 248   D. 268

15. $a + b = b + \underline{\ ?\ }$

A. a C. c

B. b D. d

16. $(a + b) + c = a + (\underline{\ ?\ } + c)$

A. a C. c

B. b D. d

17. $a \times 1 = \underline{\ ?\ }$

A. a C. c

B. b D. d

18. $\underline{\ ?\ } \times b = 0$

A. 10   C. 1

B. 5   D. 0

Mark the operation that makes the equation true.

19. $\frac{1}{3} \bigcirc \frac{1}{3} = \frac{2}{3}$

A. +   C. ×

B. −   D. ÷

20. $92 \bigcirc 8 + 25 = 125$

A. +   C. ×

B. −   D. ÷

21. $36 \bigcirc 6 = 6$

A. +   C. ×

B. −   D. ÷

22. $\$8.59 \bigcirc \$1.45 = \$7.14$

A. +   C. ×

B. −   D. ÷

Expressions

Write a mathematical expression for the number.

1. 18
    _____

2. 14
    _____

3. 9
    _____

4. 20
    _____

Write a mathematical equation for the number.

5. 18
    _____

6. 14
    _____

7. 9
    _____

8. 20
    _____

Write >, <, or = to make the statement true.

9. $3 \times 12 \bigcirc 5 + 20$

10. $50 - 8 \bigcirc 6 \times 7$

11. $68 \times 0 \bigcirc 68 + 0$

Write +, −, ×, or ÷ to make the statement true.

12. $(3 \times 5) \bigcirc 8 = 2 \times 10 + 3$

13. $3 \bigcirc 8 = 8 + 8 + 8$

14. $63 \bigcirc 9 = 17 - 10$

Write a mathematical expression.

15. Mr. George divided the candy into 20 bags. _____

16. Kristin collected 5 more than 1 dozen seashells. _____

17. Five of the 8 tomato plants died. _____

Solve. Label when applicable.

18. Lori has taken 3 history tests. Her scores are 98, 104, and 89. What is her average test score?

19. Janet loves to scrapbook. She has completed 5 albums this year. If each album has 45 pages, how many pages has she completed in all?

20. Joey bought 5 cards to add to his already large collection of 1,089 cards. How many cards does he have in all?

21. There are 12 students in Miss White's class. Each student has 7 textbooks at his desk. Miss White has 7 textbooks in addition to 7 teacher's books. How many books are there in all?

Complete the table.

	word form	eighteen million, six hundred thirty-four thousand, nine hundred seven
1.	standard form	
2.	expanded form	
3.	expanded form with multiplication	

Write the number in expanded form.

4. $5.124 = $ ___$5 + 0.1 + 0.02 + 0.004$___

5. $6.21 = $ _____

6. $31.92 = $ _____

7. $3.054 = $ _____

8. $0.891 = $ _____

9. $0.129 = $ _____

10. $9.102 = $ _____

11. $8.01 = $ _____

Write the decimal.

12. seven and five tenths = _____

13. $2\frac{67}{100} = $ _____

14. twenty-two and sixteen hundredths = _____

15. eight tenths = _____

16. four and seven thousandths = _____

17. five and thirty-six hundredths = _____

18. $\frac{51}{100} = $ _____

19. $\frac{4}{10} = $ _____

Match.

_____ 20. 8.745

_____ 21. 6.574

_____ 22. 8.754

_____ 23. 3.457

_____ 24. 7.386

A The value of 7 is seven thousandths.

B $8 + 0.7 + 0.05 + 0.004$

C $7\frac{386}{1,000}$

D The value of 6 is six ones.

E eight and seven hundred forty-five thousandths

Equations

Name _____

Write the value of *n*.

1.

100	
n	*n*

$n =$ _____

2.

n		
3	3	3

$n =$ _____

3.

n	
5	10

$n =$ _____

Write >, <, or = to make the statement true.

4. $25 \bigcirc 5 \times 2$

5. $543 - 40 \bigcirc 54 + 403$

6. $3 \times 3 \times 2 \bigcirc 2 \times 3 \times 4$

7. $5 + 3.4 \bigcirc 53 + 0.4$

8. $25 + 8 \bigcirc 8 + 25$

9. $4^2 \bigcirc 36 - 20$

10. $(10 \times 2) \times 3 \bigcirc 10 \times (2 \times 3)$

11. $\frac{4}{4} + 10 \bigcirc \frac{10}{10} + 4$

12. $4 \times 3 \bigcirc 5 \times 4$

13. $\frac{63}{9} \bigcirc 9\overline{)72}$

14. $81 \div 9 \bigcirc 3 \times 3$

15. $24 \times 5 \bigcirc (20 \times 4) + 5$

Use mental math to match the expressions of equal value.

_____ **16.** $(5 \times 2) + (2 \times 9)$ **A** $(12 \times 2) - 1$

_____ **17.** $(3 \times 5) + 8$ **B** $3{,}590 + 12$

_____ **18.** $3{,}000 + 600 + 2$ **C** $84 \div 7$

_____ **19.** $3\frac{1}{4} + 8\frac{3}{4}$ **D** 7×4

Draw a picture. Solve. Label your answer.

20. Brian had 28 songs downloaded on his computer. He deleted 7 of them. How many does he have left?

21. Mrs. Thompson bought 8 pizzas for the class Christmas party. Each pizza was cut into 8 slices. The class ate $5\frac{3}{8}$ of the pizzas. How many slices did they eat?

Use the given operation to write a mathematical expression for the value.

1.

15
_____ + _____
_____ − _____
_____ × _____
_____ ÷ _____

2.

28
_____ + _____
_____ − _____
_____ × _____
_____ ÷ _____

3.

42
_____ + _____
_____ − _____
_____ × _____
_____ ÷ _____

4.

99
_____ + _____
_____ − _____
_____ × _____
_____ ÷ _____

Write a mathematical expression.

5. Joe had twice as many baseball cards as Dan. _____

6. the difference between the boiling point and the freezing point of water (°F) _____

7. The result was 14 points less than 100. _____

8. It rained twice as much this year as it did last year. _____

Write the opposite number.

9. 5 _____

10. 5,321 _____

11. ⁻43 _____

12. 98 _____

13. ⁻214 _____

14. 2 _____

15. ⁻105 _____

16. ⁻391 _____

Write > or < to compare.

17. 5 ◯ ⁻5

18. ⁻4 ◯ 3

19. 8 ◯ ⁻20

20. 5 ◯ ⁻15

Write the numbers from *least* to *greatest*.

21.

⁻6	4	⁻15

_____ _____ _____

22.

17	3	⁻20

_____ _____ _____

23.

⁻10	⁻4	⁻21

_____ _____ _____

Solve.

24.
$$\begin{array}{r} 77 \\ \times\ 5 \\ \hline \end{array}$$

25.
$$\begin{array}{r} 671 \\ \times\ 5 \\ \hline \end{array}$$

26.
$$\begin{array}{r} \$7.29 \\ \times\ 61 \\ \hline \end{array}$$

27.
$$\begin{array}{r} 507 \\ \times\ 56 \\ \hline \end{array}$$

28.
$$\begin{array}{r} 25 \\ \times 70 \\ \hline \end{array}$$

29. 3 • 7 = _____

30. 5 • 4 = _____

31. 8 • 9 = _____

32. 8 • 6 = _____

33. 9 • 9 = _____

Balanced Equations

Name _____

Write the value of the object.

1.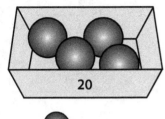

20

⬤ = _____

2.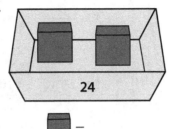

24

▮ = _____

3.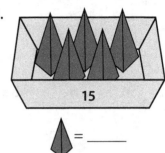

15

◢ = _____

Write the total value of the objects. Write >, <, or = to compare the total values.

 = 5   = 2   ▮ = 4

4.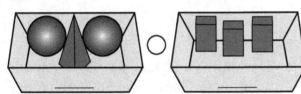

_____   ◯   _____

5.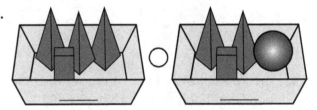

_____   ◯   _____

Write the value of the expression if $n = 6$.

6. $(4 \times n) + 5 =$ _____

7. $90 - (n \cdot 12) =$ _____

8. $150 \div n =$ _____

9. $(n \cdot 6) + 36 =$ _____

10. $5 \cdot n =$ _____

11. $(n + 94) \div 10 =$ _____

12. $n^2 + 4 =$ _____

13. $(54 \div n) - 9 =$ _____

14. $(48 \div n) \cdot 8 =$ _____

15. $11 \cdot n =$ _____

16. $n + n =$ _____

17. $(58 - n) \cdot 2 =$ _____

Write the value of n.

18. $4 + n = 10$

$n =$ _____

19. $6 \cdot 4 = 29 - n$

$n =$ _____

20. $n \times 3 = 48$

$n =$ _____

21. $n + 2 = 4 \cdot 6$

$n =$ _____

22. $36 \div n = 12$

$n =$ _____

23. $150 - n = 2 \cdot 25$

$n =$ _____

24. $6 \cdot 4 = 8 \cdot n$

$n =$ _____

25. $n^2 = 49$

$n =$ _____

Write a mathematical expression.

26. Ann worked 3 hours more than Hal. _____

27. Mrs. Cox made 2 cookies for each student in her class. _____

Write the value of *n*.

1.

n	
32	68

n = _____

2.

500	
200	*n*

n = _____

3.

16	
n	*n*

n = _____

4.

n		
5	5	5

n = _____

5.

n			
10	10	10	10

n = _____

6.

24		
n	*n*	*n*

n = _____

Solve. Label your answer.

7. Jill has 23 quarters for her state quarter coin collection. The full set includes 56 quarters, one for each US state and territory. How many does she still need to collect?

8. Mount Everest is 29,035 feet high. Denali is 20,320 feet high. How much higher is Mount Everest than Denali?

9. The Poole family keeps a record of their spending. Each month they pay $950 for housing, $80 for electricity, $110 for internet and cable, and $120 for homeowner's insurance. How much do they spend every month?

10. The Smiths budgeted $4,500 to remodel their home. They spent $800 on paint, $300 on tools, and $1,500 on building materials. How much money do they have left for furniture?

Use a ruler to draw the figure. Name the figure using symbols.

11. **line**

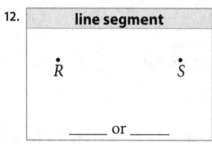

_____ or _____

12. **line segment**

_____ or _____

13. **line**

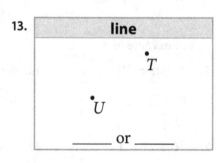

_____ or _____

Use plane *x* to find the answer. Write the answer using symbols.

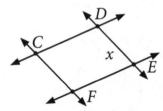

14. Name a line segment. _____

15. Name a point on the plane. _____

16. Name a line. _____

Equations in Word Problems

Name _____

Write a mathematical equation for the problem. Solve. Label your answer.

1. Miss Swift purchased some new items for her classroom. She spent $99.50 for a new microwave, $45.99 for a month's supply of candy, and $10.50 for stickers. How much did she spend in all?

2. Mr. Muller had $500 to spend for new lawn equipment. He bought a lawnmower for $169.95. How much money does he have left?

3. Mr. LeBlanc sponsored a summer mission trip out west. The team traveled 2,200 miles in 4 days. What is the average number of miles that they traveled per day?

4. Mrs. Clater purchased 5 stickers and 4 pencils for each of her 25 students. How many items did she purchase in all?

Use the part-part-whole model to solve. Label your answer.

5. Christine practices her violin 5 days per week. She practiced 2 hours and 30 minutes for the week. How many minutes did she spend practicing each day?

2 hours 30 minutes				
_____ minutes				

6. The Red Cross collected a total of 149 quarts and 2 pints of blood in 4 days. How many pints of blood did they collect each day?

149 quarts 2 pints			
_____ pints			

Solve for n.

7.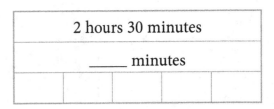

2 qt		
1 qt	1 pt	n

n = _____

8.

5 lb			
2 lb	1 lb	16 oz	8 oz	n

n = _____

9.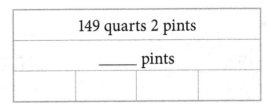

3 ft		
2 ft	6 in.	n

n = _____

10.

1 hr		
30 min	15 min	n

n = _____

Write the value of the object.

1.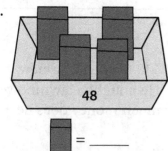

48

☐ = _____

2.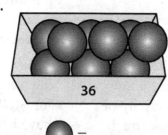

36

⬤ = _____

3.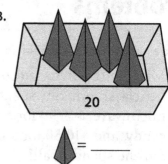

20

◭ = _____

Write a mathematical expression.

4. Jim's driveway is twice as long as Brian's. _____

5. Katie was in Oahu for 14 days and in Lanai for 7 days. _____

6. Mrs. Harding passed out 5 candies to each of her 25 students. _____

Complete the table.

7.

Rule: ÷ 12	
Input	Output
12	
48	
108	
180	
600	

8.

Rule: ÷ 5	
Input	Output
10	
105	
220	
445	
905	

9.

Rule: ÷ 10	
Input	Output
40	
100	
1,000	
1,540	
2,500	

10.

Rule: ÷ 3	
Input	Output
6	
300	
636	
933	
1,050	

Solve.

11.
$$23\overline{)862}$$

12.
$$52\overline{)1,493}$$

13.
$$2\overline{)4,496}$$

Rename the fraction to lowest terms. Show your work.

14. $\frac{12}{14} =$ ☐

Solve

$\frac{12 \div 2}{14 \div 2}$

15. $\frac{33}{60} =$ ☐

Solve

16. $\frac{14}{28} =$ ☐

Solve

17. $\frac{3}{15} =$ ☐

Solve

Equations Review

Mark the equivalent expression.
Mark *NH* if the answer is "Not Here."

1. 24
   - ○ $(10 + 2) - 4$
   - ○ $23 + 1$
   - ○ 8×2
   - ○ NH

2. 10×12
   - ○ $(10 \times 10) + 2$
   - ○ $12 \times (5 + 2)$
   - ○ $100 + 20$
   - ○ NH

3. $32 - 10$
   - ○ $(30 + 2) \times 10$
   - ○ 11×3
   - ○ $20 + 10$
   - ○ NH

4. $(10 + 5) - 5$
   - ○ 5×2
   - ○ $15 \div 2$
   - ○ $10 + 5^2$
   - ○ NH

5. $63 \div 7$
   - ○ $60 \div 10$
   - ○ $60 + (3 \times 7)$
   - ○ $81 \div 9$
   - ○ NH

Write the correct number to make the expression equal the given value.

6.

24
_____ + 5
34 − _____
_____ × 8
48 ÷ _____

Mark the value of the object.

7.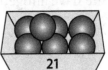
   21
    $= \underline{\ ?\ }$
   - ○ 2
   - ○ 3
   - ○ 7
   - ○ NH

8.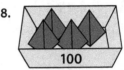
   100
    $= \underline{\ ?\ }$
   - ○ 10
   - ○ 25
   - ○ 50
   - ○ NH

9.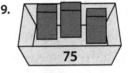
   75
    $= \underline{\ ?\ }$
   - ○ 20
   - ○ 25
   - ○ 30
   - ○ NH

Mark the value of *n*.

10. $6 \cdot 4 = 8 \cdot n$
    - ○ 1
    - ○ 2
    - ○ 3
    - ○ NH

11. $51 - 11 = n$
    - ○ 20
    - ○ 40
    - ○ 62
    - ○ NH

12. $10 - n = 21 \div 7$
    - ○ 7
    - ○ 8
    - ○ 9
    - ○ NH

13. $n + 29 = 47$
    - ○ 18
    - ○ 12
    - ○ 14
    - ○ NH

Mark the equation that solves the problem.

14. Moriah bought 3 bags of hard candy. There were 24 pieces of candy in each bag. How many pieces did she have altogether?

    ○   ○   ○

    $24 \times 3 = 72$ $24 \div 3 = 8$ $24 + 3 = 27$

15. Jason's new book has 450 pages and 19 chapters. He read the book in 5 days. What is the average number of pages he read each day?

    ○   ○   ○

    $450 \div 18 = 25$ $19 \times 5 = 95$ $450 \div 5 = 90$

16. Mrs. Smith is making 3 dresses for her daughter's wedding. Each dress needs 5 yards of fabric. How much fabric will she need to purchase in all?

    ○   ○   ○

    $5 + 3 = 8$ $5 \times 3 = 15$ $5 \div 3 = 1.8$

Write a mathematical expression.

17. Lori earned $254 and spent $103.

    _____

18. Matt exercised 30 minutes a day for 1 week.

    _____

19. The truck used 24 gallons of gas in 6 hours.

    _____

Write the value of *n* for the part-part-whole model.

20.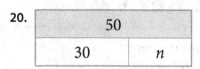

50	
30	n

$n =$ _____

21.

24		
n	n	n

$n =$ _____

22.

n		
6	6	6

$n =$ _____

23.

8			
2	2	2	n

$n =$ _____

Write the value of the expression if *n* = 5.

24. $n + 4$

    _____

25. $(7 - n) \cdot 50$

    _____

26. $155 \div n$

    _____

27. $(n \cdot 6) \div 5$

    _____

Write the value of *n*.

28. $(6 \cdot n) = 150 \div 5$

    _____

    _____

Name _____

Mark the answer.

1. $70\overline{)5,600}$

   A. 80   C. 90
   B. 800   D. 900

2. 30×90

   A. 270   C. 280
   B. 2,700 D. 2,800

3. $600 + 700$

   A. 120   C. 130
   B. 1,200 D. 1,300

4. $2,700 - 800$

   A. 2,100 C. 1,700
   B. 1,900 D. 1,500

5. $30\overline{)9,060}$

   A. 302   C. 32
   B. 320   D. 3,002

6. 221×4

   A. 800   C. 884
   B. 880   D. 888

7. $41.5 + 48.05$

   A. 89.55 C. 90.055
   B. 89.055 D. 90.55

8.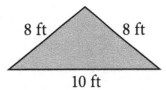

   The perimeter is __?__.
   A. 640 ft   C. 36 ft
   B. 160 ft   D. 26 ft

9.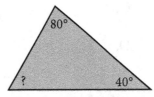

   The missing angle measures __?__.
   A. 60°   C. 100°
   B. 90°   D. 120°

10.

    The 4 angles are __?__.
    A. acute   C. right
    B. obtuse D. straight

11.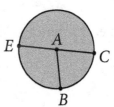

    Chord *EC* is also __?__.
    A. the diameter   C. a radius
    B. the name of the circle D. the vertex

Mark the equivalent fraction.

12.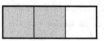

$\frac{2}{3} = \frac{?}{\underline{\quad}}$

A. $\frac{3}{4}$ C. $\frac{4}{6}$

B. $\frac{4}{9}$ D. $\frac{4}{12}$

13.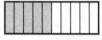

$\frac{5}{10} = \frac{?}{\underline{\quad}}$

A. $\frac{2}{5}$ C. $\frac{2}{3}$

B. $\frac{1}{2}$ D. $\frac{5}{5}$

14.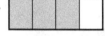

$\frac{3}{4} = \frac{?}{\underline{\quad}}$

A. $\frac{1}{2}$ C. $\frac{6}{8}$

B. $\frac{4}{4}$ D. $\frac{5}{10}$

Mark the true statement.

15.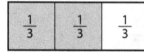

A. The shaded part is greater than $\frac{1}{2}$.

B. The shaded part is less than $\frac{1}{2}$.

C. The shaded part is equivalent to $\frac{4}{8}$.

16.

| $\frac{1}{8}$ | $\frac{1}{8}$ | $\frac{1}{8}$ | $\frac{1}{8}$ | $\frac{1}{8}$ | $\frac{1}{8}$ | $\frac{1}{8}$ | $\frac{1}{8}$ |

A. The shaded part is equal to $\frac{1}{2}$.

B. The shaded part is greater than 1 whole.

C. The shaded part is greater than $\frac{1}{2}$.

Mark the sum.

17. $\frac{1}{2} + \frac{1}{4} = \frac{?}{\underline{\quad}}$

A. $\frac{2}{4}$ C. $\frac{3}{2}$

B. $\frac{3}{4}$ D. 1

18. $\frac{2}{3} + \frac{1}{9} = \frac{?}{\underline{\quad}}$

A. 1   C. $\frac{7}{9}$

B. $\frac{3}{9}$ D. $1\frac{1}{9}$

19. $\frac{1}{2} + \frac{1}{2} = \frac{?}{\underline{\quad}}$

A. $\frac{2}{4}$ C. $\frac{1}{2}$

B. 1   D. $\frac{3}{4}$

20. $\frac{2}{4} + \frac{2}{3} = \frac{?}{\underline{\quad}}$

A. $1\frac{1}{6}$ C. $\frac{11}{12}$

B. $1\frac{2}{3}$ D. $\frac{4}{12}$

21. $\frac{8}{8} + \frac{1}{7} = \frac{?}{\underline{\quad}}$

A. $8\frac{1}{7}$ C. $1\frac{1}{7}$

B. $1\frac{7}{8}$ D. $8\frac{1}{8}$

22. $\frac{1}{6} + \frac{1}{4} = \frac{?}{\underline{\quad}}$

A. $\frac{1}{12}$ C. $\frac{3}{10}$

B. $\frac{5}{12}$ D. $\frac{2}{6}$

Quadrilaterals & Other Polygons

Name _____

A **polygon** is a closed figure made of three or more line segments.

Mark the figure as a polygon or not a polygon.

1.
   ○ polygon
   ○ not a polygon

2.
   ○ polygon
   ○ not a polygon

3.
   ○ polygon
   ○ not a polygon

4.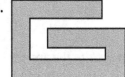
   ○ polygon
   ○ not a polygon

Write the number of sides. Name the polygon.

5.
   _____ sides

   _____

6.
   _____ sides

   _____

7.
   _____ sides

   _____

heptagon
hexagon
octagon
pentagon
quadrilateral

8.
   _____ sides

   _____

9.
   _____ sides

   _____

10.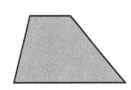
   _____ sides

   _____

Write the measurement of the unknown angle.

11.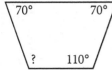
   _____

12.
   | 95° | 85° |
   | ? | 95° |
   _____

A **quadrilateral** is a polygon with 4 sides. The sum of the angles measures 360°.

$$\angle A + \angle B + \angle C + \angle D = 360°$$

13.
   _____

Write a mathematical expression.

1. Matt gave 8 of his 15 baseball cards to Ian. _____

2. Joel bought 5 boxes of 40 drinks each. _____

3. Twice as much grass seed was needed for the yard. _____

4. 50 flyers were distributed among 25 doors. _____

5. Dora made 2 cookies for each of the 53 teachers. _____

Add. Write the answer in lowest terms.

6. $\dfrac{5}{10}$ $+ \dfrac{3}{10}$

7. $\dfrac{8}{18}$ $+ \dfrac{7}{18}$

8. $\dfrac{1}{8}$ $+ \dfrac{5}{8}$

9. $\dfrac{8}{10}$ $+ \dfrac{3}{10}$

10. $\dfrac{3}{4}$ $+ \dfrac{1}{4}$

11. $\dfrac{4}{5}$ $+ \dfrac{5}{5}$

12. $3\dfrac{1}{2}$ $+ 4\dfrac{1}{2}$

13. $\dfrac{4}{9}$ $+ \dfrac{7}{9}$

14. $2\dfrac{1}{6}$ $+ 1\dfrac{5}{6}$

15. $6\dfrac{5}{8}$ $+ 3\dfrac{7}{8}$

Write the product.

16. $50 \times 20 =$ _____

17. $20 \times 40 =$ _____

18. $30 \times 5 =$ _____

19. $70 \times 8 =$ _____

20. $80 \times 900 =$ _____

21. $70 \times 60 =$ _____

22. $30 \times 120 =$ _____

23. $40 \times 8 =$ _____

24. $70 \times 70 =$ _____

Use the Distributive Property to find the products mentally.

25. $6 \times 22 =$ _____

26. $7 \times 16 =$ _____

27. $3 \times 55 =$ _____

28. $5 \times 18 =$ _____

29. $8 \times 25 =$ _____

30. $6 \times 24 =$ _____

Solve.

31. $33 \div 3 =$ _____

32. $16 - 4 =$ _____

33. $8 \times 4 =$ _____

34. $15 + 8 =$ _____

35. $21 - 7 =$ _____

36. $5 \times 9 =$ _____

37. $\dfrac{42}{6} =$ _____

38. $24 + 6 =$ _____

Perimeter & Circumference

Perimeter is the distance around a figure.

3 m

$P = s + s + s + s$
$P = 3\,m + 3\,m + 3\,m + 3\,m$
or
$4 \times 3\,m = 12\,m$

Label the polygon *regular* or *irregular*. Find the perimeter of the figure. Solve and label.

1.
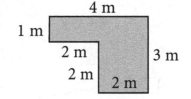

4 m
1 m
2 m
2 m
3 m
2 m

_____

$P =$ _____

_____

2.

5 in.

_____

$P =$ _____

_____

Circumference is the distance around a circle.

$C = 3.14 \times diameter$

Estimate the circumference.

8 in.

$C = \boxed{3} \times 8\,in. = 24\,in.$

The distance around the circle is about 24 inches.

Solve. Label your answer.

3. Mary wants to put lace around a small round pillow. The diameter of the pillow is 11 inches. About how much lace will she need?

diameter = _____

circumference = ___ 3 ___ × _____

Mary needs about _____ of lace.

4. Tim has a basketball with a circumference of 30 inches. Tim's brother has a basketball hoop with a diameter of 9 inches. Estimate the circumference of the hoop.

___ 3 ___ × _____ = _____

Will the basketball fit through the hoop? Why or why not?

_____

Compare the figures. Write the symbol for similar (∼) or congruent (≅).

5.

6.

7.
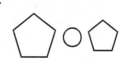

8.

$\overline{OP}$ ○ $\overline{QR}$

Write *translation*, *rotation*, or *reflection* to show how the figure moved.

9.
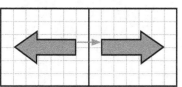

_____

10.
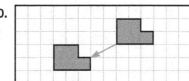

_____

11.
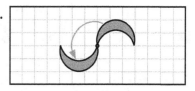

_____

Use the equation $\angle A + \angle B + \angle C + \angle D = 360°$ **to find the measurement of the unknown angle.**

1.

   _____

2.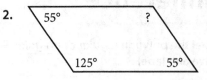

   _____

3.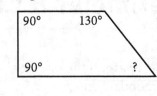

   _____

4.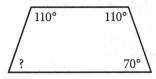

   _____

5.
   _____

6.
   _____

Add. Write the answer in lowest terms.

7. $\dfrac{1}{3}$
    $+\ \dfrac{5}{6}$

8. $\dfrac{2}{4}$
    $+\ \dfrac{1}{8}$

9. $\dfrac{2}{10}$
    $+\ \dfrac{6}{10}$

10. $\dfrac{2}{9}$
    $+\ \dfrac{5}{3}$

11. $1\dfrac{5}{8}$
    $+\ 3\dfrac{2}{4}$

12. $2\dfrac{1}{8}$
    $+\ 4\dfrac{3}{4}$

13. $1\dfrac{2}{5}$
    $+\ 7\dfrac{1}{10}$

14. $6\dfrac{3}{4}$
    $+\ 1\dfrac{1}{8}$

Fill in the missing digits.

15.
```
  4 8,☐5 2
- 2 ☐,0 8 4
───────────
  2 8,5 6 8
```

16.
```
  9 3,☐4 3
- 5 1,4 9 8
───────────
  4 2,0 4 ☐
```

17.
```
  ☐,0 8 7
- 4,1 5 8
─────────
    9 2 ☐
```

18.
```
  8 6,3 8 ☐
- ☐,2 4 2
───────────
  8 4,1 4 0
```

Solve.

19. $9 \times 7 =$ _____

20. $10 + 15 =$ _____

21. $7\overline{)28}$

22. $16 - 7 =$ _____

23. $20 + 25 =$ _____

24. $\dfrac{56}{8} =$ _____

25. $7 \cdot 6 =$ _____

26. $\dfrac{21}{7} =$ _____

27. $4 \cdot 8 =$ _____

28. $20 - 10 =$ _____

Classifying Triangles

Name _____

acute triangle	**right triangle**	**obtuse triangle**
all angles less than 90°	one 90° angle	one angle greater than 90°

Classify the triangle as *acute*, *obtuse*, or *right*. Find the measure of the unknown angle.

1. 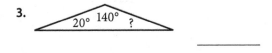   _____

$180° - ($ _____ $+$ _____ $) =$ ___?___

$180° - $ _____ $=$ _____

2. 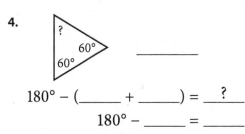   _____

$180° - ($ _____ $+$ _____ $) =$ ___?___

$180° - $ _____ $=$ _____

3. 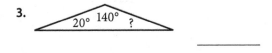   _____

$180° - ($ _____ $+$ _____ $) =$ ___?___

$180° - $ _____ $=$ _____

4. 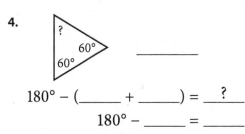   _____

$180° - ($ _____ $+$ _____ $) =$ ___?___

$180° - $ _____ $=$ _____

equilateral triangle	**isosceles triangle**	**scalene triangle**
all sides equal	2 equal sides	no equal sides

Classify the triangle according to the lengths of its sides. Find the perimeter of the triangle.

5.

_____

$P =$ __2 ft__ $+$ _____ $+$ _____

$P =$ _____

6.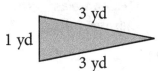

_____

$P =$ _____ $+$ _____ $+$ _____

$P =$ _____

7.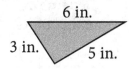

_____

$P =$ _____ $+$ _____ $+$ _____

$P =$ _____

Solve. Label when applicable.

8. A pennant measures 3 ft × 6 ft × 6 ft. What is the perimeter?

_____

What type of triangle is formed by the sides? _____

The angles of the pennant measure 80°, 80°, and 20°. Is the triangle acute, obtuse, or right? _____

9. The yield sign at the end of the street measures 2 ft × 2 ft × 2 ft. What is the perimeter of the sign?

_____

What type of triangle is formed by the sides? _____

All three angles measure 60°. Is the triangle acute, obtuse, or right? _____

Estimate the circumference of the circle.

1.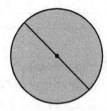

diameter = 5 in.

$C =$ _____ $\times$ _____ $=$ _____

2.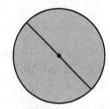

diameter = 12 ft

$C =$ _____

3.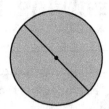

diameter = 25 cm

$C =$ _____

Add or subtract. Write the answer in lowest terms.

4. $4\frac{3}{4}$
$-\ 1\frac{1}{4}$

5. $5\frac{2}{3}$
$-\ 2\frac{1}{15}$

6. $3\frac{2}{3}$
$+\ 1\frac{4}{15}$

7. $3\frac{1}{5}$
$+\ 1\frac{9}{10}$

8. $6\frac{11}{12}$
$-\ 4\frac{5}{6}$

9. $4\frac{5}{12}$
$+\ 1\frac{1}{3}$

10. $2\frac{6}{9}$
$-\ \ \frac{2}{3}$

11. $3\frac{5}{14}$
$-\ 1\frac{2}{7}$

Solve. Rename the units.

	Rename as smaller units.			Rename as larger units.			Rename as 2 units.	
12.	2 tn = _____ lb	17.	6 pt = _____ qt	22.	41 oz = _____ lb _____ oz
13.	5 gal = _____ qt	18.	64 oz = _____ lb	23.	17 c = _____ pt _____ c
14.	4 yd = _____ ft	19.	36 in. = _____ yd	24.	33 hr = _____ day _____ hr
15.	3 hr = _____ min	20.	28 days = _____ wk	25.	50 in. = _____ ft _____ in.
16.	3 days = _____ hr	21.	120 min = _____ hr	26.	9 qt = _____ gal _____ qt

Solve.

27. $70 \div 7 =$ _____

28. $18 - 6 =$ _____

29. $9 + 5 =$ _____

30. $11 \times 10 =$ _____

31. $108 \div 12 =$ _____

32. $12 - 5 =$ _____

33. $8 + 6 =$ _____

34. $9 \times 7 =$ _____

Area

Name _____

Write a multiplication equation to find the area. Solve.

1.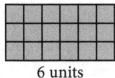

   3 units

   6 units

   ____ units × ____ units =

   ____ units²

2.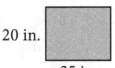

   20 in.

   25 in.

   ____ in. × ____ in. =

   ____ in.²

3.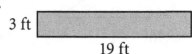

   3 ft

   19 ft

   _____

   _____

4.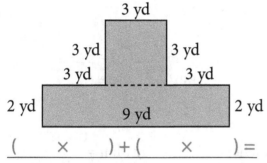

   3 yd

   3 yd   3 yd

   3 yd   3 yd

   2 yd   9 yd   2 yd

   (____ × ____) + (____ × ____) =

   _____

5.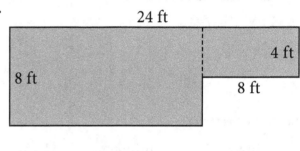

   24 ft

   8 ft   4 ft

   8 ft

   _____

   _____

Use the chart to answer the question.

Mr. Barbrow needs to purchase carpet for the elementary school.

Room	Length	Width
office	10 ft	20 ft
clinic	12 ft	8 ft
hallway	4 ft	25 ft
music room	20 ft	30 ft

6. What is the area of the office and the clinic?

   _____

7. What is the area of the hallway and music room?

   _____

Name the shape according to the number of sides. Write an equation to find the perimeter.

8.

   2 in.

   _____

   _____

9.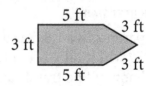

   5 ft   3 ft

   3 ft

   5 ft   3 ft

   _____

   _____

10. Each side of a quadrilateral measures 6 inches. It has 4 right angles.

    _____

    _____

Classify the triangle as *acute*, *obtuse*, or *right*. Find the unknown angle.

1.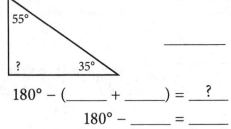

   _____

   $180° − ($_____$+$_____$) =$ ___?___

   $180° −$_____$=$_____

2.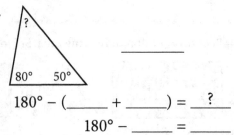

   _____

   $180° − ($_____$+$_____$) =$ ___?___

   $180° −$_____$=$_____

3.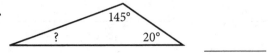

   _____

   $180° − ($_____$+$_____$) =$ ___?___

   $180° −$_____$=$_____

4.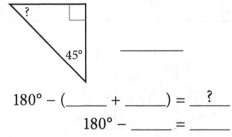

   _____

   $180° − ($_____$+$_____$) =$ ___?___

   $180° −$_____$=$_____

Complete the facts.

5. 1 hour = _____ minutes

6. 1 yard = _____ inches

7. 1 gallon = _____ quarts

8. 1 year = _____ weeks

9. 1 minute = _____ seconds

10. 1 ton = _____ pounds

11. 1 day = _____ hours

12. 1 year = _____ months

13. 1 quart = _____ pints

14. 1 foot = _____ inches

15. 1 pound = _____ ounces

16. 1 yard = _____ feet

Divide.

17.
$$24)\overline{4,782}$$

18.
$$50)\overline{6,913}$$

19.
$$31)\overline{836}$$

Solve. Label your answer.

20. Brian's parents live 1,885 miles away. How long will it take Brian to reach his parents' home if he travels an average of 65 miles per hour?

21. Doris is making monster chocolate chip cookies for Teacher Appreciation Day. She has 2,520 chocolate chips. If she puts 30 chips in each cookie, how many cookies can she make?

Solve.

22. $\frac{21}{7} =$ _____

23. $12 \cdot 6 =$ _____

24. $14 + 6 =$ _____

25. $14 − 5 =$ _____

Area of a Triangle

Name _____

Divide the area of a rectangle by 2 to get the area of the related triangle.

Write an equation to find the area of the shaded part. Solve.

1.

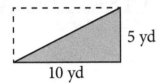

5 yd
10 yd

A of ☐ = _____

A of △ = _____

2.

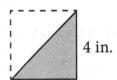

4 in.

A of ☐ = _____

A of △ = _____

3.

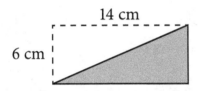

14 cm
6 cm

A of ☐ = _____

A of △ = _____

4.

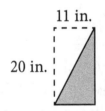

11 in.
20 in.

A of ☐ = _____

A of △ = _____

Write an equation to find the area.

5.

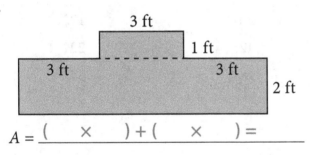

3 ft
1 ft
3 ft   3 ft
2 ft

A = (___ × ___) + (___ × ___) = _____

A = _____

6.

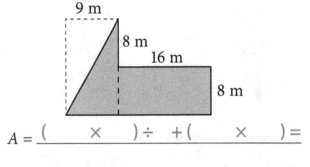

9 m
8 m
16 m
8 m

A = (___ × ___) ÷ ___ + (___ × ___) = _____

A = _____

Solve. Label your answer.

7. Mr. Carruthers installed hardwood floors in his living room and hallway. How many square feet of hardwood did he need?

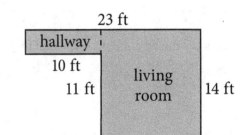

23 ft
hallway
10 ft
11 ft
living room
14 ft

_____

He also put new baseboards all along the walls. How many feet of baseboard did he need?

_____

Find the area.

1. Barbara just picked out laminate flooring for her house. How many square feet will she need to purchase?

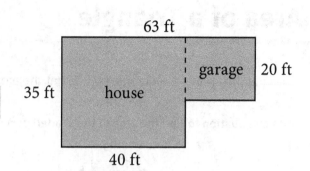

Classify the triangle according to the lengths of its sides.

equilateral	isosceles	scalene

2.

1 in.

_____

3.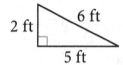

2 ft, 6 ft, 5 ft

_____

4.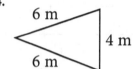

6 m, 4 m, 6 m

_____

5.

6 yd

_____

Complete the table.

6.

Rule: ÷ 3	
Input	Output
18	
84	
192	
246	
948	

7.

Rule: × 7	
Input	Output
5	
10	
25	
100	
150	

8.

Rule: + 25	
Input	Output
20	
115	
231	
425	
980	

9.

Rule: − 9	
Input	Output
23	
69	
102	
221	
543	

Write >, <, or = to compare.

10. $\frac{3}{6}$ ◯ $\frac{3}{4}$

11. $\frac{10}{12}$ ◯ $\frac{4}{24}$

12. $\frac{6}{8}$ ◯ $\frac{14}{8}$

13. $\frac{3}{10}$ ◯ $\frac{1}{5}$

14. $2\frac{4}{5}$ ◯ $2\frac{8}{10}$

15. $3\frac{2}{12}$ ◯ $3\frac{1}{4}$

16. $5\frac{6}{10}$ ◯ $5\frac{4}{14}$

17. $\frac{2}{3}$ ◯ $\frac{8}{12}$

Write the mixed numbers from _least_ to _greatest_.

18. $4\frac{1}{2}$ $3\frac{6}{8}$ $3\frac{5}{16}$

19. $2\frac{4}{6}$ $2\frac{1}{2}$ $2\frac{5}{6}$

20. $4\frac{8}{10}$ $11\frac{3}{10}$ $4\frac{2}{5}$

_____ _____ _____

Math 5 Activities

Perimeter & Area

Name _____

Use the diagram below to solve the problem. Write the equation you used.

1. The yard crew will seed half of the playground this summer. What is the area that needs to be covered?

2. The other half of the playground will be covered with mulch. A truckload of mulch covers about 162 square feet. How many loads are needed?

3. Each load of mulch costs $25.75. How much will be spent on mulch?

4. A fence was built around the playground. What is the perimeter of the playground?

5. Mr. Vick replaced the floor of the cafeteria with new ceramic tile. What is the area of the cafeteria?

6. The front and back halls and classrooms need new carpet. What is the area of the front and back halls and classrooms?

7. New smoke detectors were installed every 30 feet along the outer wall of the front and back halls and along one side of the main hall. How many smoke detectors were needed?

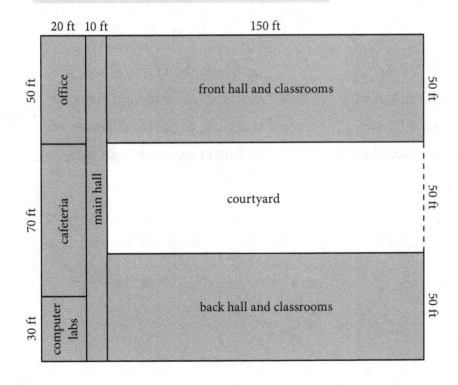

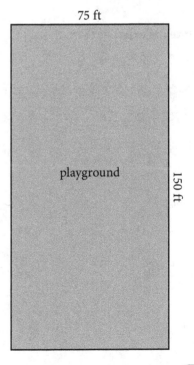

Write an equation to find the area of the shaded part.

1.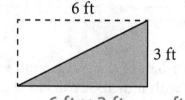
6 ft

3 ft

A of ▭ = __6 ft × 3 ft =___ ft²__
A of ▲ = __18 ft² ÷ 2 =___ ft²__

2.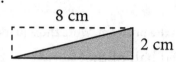
8 cm

2 cm

Area of a triangle = $\frac{1}{2}$ the area of the related rectangle.

A of ▭ = _____
A of ▲ = _____

3.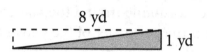
8 yd

1 yd

A of ▭ = _____
A of ▲ = _____

4.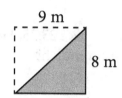
9 m

8 m

A of ▭ = _____
A of ▲ = _____

Write the factors of the number. Label the number _prime_ or _composite_.

5. 5: _____

6. 18: _____

7. 24: _____

8. 12: _____

9. 37: _____

10. 42: _____

Write the exponent form.

11. $7 \times 7 \times 7 =$ _____

12. $10 =$ _____

13. five to the fourth power = _____

14. $9 \times 9 \times 9 \times 9 \times 9 =$ _____

15. $1,000,000 =$ _____

16. three to the second power = _____

17. $6 =$ _____

18. $1,000 =$ _____

19. eleven to the third power = _____

20. $2 \times 2 =$ _____

21. $10,000 =$ _____

22. one to the sixth power = _____

Solve.

23. $4 + 19 =$ _____

24. $5 \times 7 =$ _____

25. $10 + 10 =$ _____

26. $3 \cdot 8 =$ _____

27. $2\overline{)26}$

28. $20 - 5 =$ _____

29. $48 \div 8 =$ _____

30. $14 - 8 =$ _____

Geometry: Perimeter & Area Review

Name _____

Mark the answer.
Mark *NH* if the answer is "Not Here."

1.
   - ○ rectangle
   - ○ rhombus
   - ○ square
   - ○ NH

2.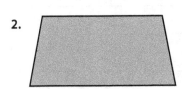
   - ○ parallelogram
   - ○ square
   - ○ rectangle
   - ○ NH

3.
   - ○ trapezoid
   - ○ parallelogram
   - ○ square
   - ○ NH

4.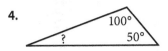
   - ○ 20°
   - ○ 30°
   - ○ 40°
   - ○ NH

5.
   - ○ 50°
   - ○ 70°
   - ○ 80°
   - ○ NH

6.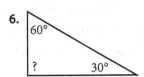
   - ○ 30°
   - ○ 60°
   - ○ 90°
   - ○ NH

7.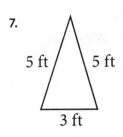
   - ○ isosceles
   - ○ scalene
   - ○ equilateral

8.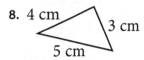
   - ○ isosceles
   - ○ scalene
   - ○ equilateral

9.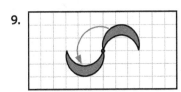
   - ○ translation
   - ○ rotation
   - ○ reflection

10.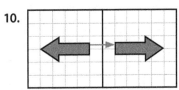
    - ○ translation
    - ○ rotation
    - ○ reflection

11.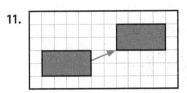
    - ○ translation
    - ○ rotation
    - ○ reflection

Match.

A similar   **B** congruent

_____ 12.

_____ 13.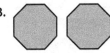

_____ 14. $\overline{WX}$ and $\overline{ZY}$

_____ 15.

_____ 16.

Match.

_____ 17.

    A heptagon

    B hexagon

    C octagon

_____ 18.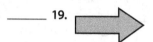

    D pentagon

    E quadrilateral

_____ 19.

_____ 20.

_____ 21.

Find the perimeter and the area.

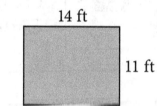

14 ft

11 ft

22. $P =$ _____

_____

23. $A =$ _____

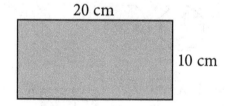

20 cm

10 cm

24. $P =$ _____

_____

25. $A =$ _____

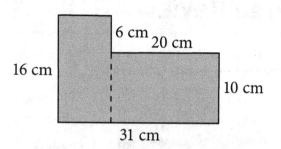

6 cm

20 cm

16 cm

10 cm

31 cm

26. $P =$ _____

_____

_____

27. $A =$ _____

_____

_____

Estimate the circumference.

28.

$C =$ _____

diameter = 9 cm

29.

$C =$ _____

diameter = 6 in.

Solve. Label your answer.

30. Mr. White had a roofer install a new roof on his house. The roof measures 56 feet by 38 feet. What is the area of the roof?

31. The cost for a new roof is $4.00 per square foot. How much did Mr. White pay?

Mark the answer.

1. $\frac{3}{9} + \frac{3}{9} = \underline{\ ?\ }$

   A. $\frac{1}{3}$ C. $\frac{3}{9}$

   B. $\frac{2}{3}$ D. $\frac{6}{18}$

2. $2 - \frac{1}{4} = \underline{\ ?\ }$

   A. $2\frac{1}{4}$ C. 2

   B. $1\frac{3}{4}$ D. $\frac{3}{4}$

3. $\begin{array}{r}\frac{5}{6}\\[-2pt]+\frac{1}{3}\\\hline\end{array}$

   A. $\frac{6}{6}$ C. $1\frac{1}{6}$

   B. $\frac{6}{3}$ D. $1\frac{4}{6}$

4. $\begin{array}{r}\frac{7}{10}\\[-2pt]-\frac{3}{5}\\\hline\end{array}$

   A. $\frac{1}{10}$ C. $\frac{4}{5}$

   B. $\frac{4}{10}$ D. $1\frac{1}{5}$

5. $\frac{1}{8} + \frac{3}{8} + \frac{2}{8} = \underline{\ ?\ }$

   A. $\frac{8}{8}$ C. $\frac{6}{6}$

   B. $\frac{3}{4}$ D. $\frac{3}{8}$

6. $\begin{array}{r}2\frac{2}{4}\\[-2pt]+2\frac{3}{4}\\\hline\end{array}$

   A. 5 C. $5\frac{3}{4}$

   B. 6 D. $5\frac{1}{4}$

7. $3 \text{ yd} = \underline{\ ?\ }$

   A. 35 in.   C. 9 ft

   B. 3 ft   D. 100 in.

8. $1 \text{ mi} = \underline{\ ?\ }$

   A. 1,000 yd   C. 12,000 in.

   B. 5,280 ft   D. 500 sq ft

9. $2 \text{ ft } 6 \text{ in.} = \underline{\ ?\ }$

   A. 26 in.   C. 3 ft

   B. 1 yd   D. 30 in.

10. $2 \text{ days} = \underline{\ ?\ }$

    A. 50 hours   C. 46 hours

    B. 48 hours   D. 42 hours

11. $5 \text{ lb} = \underline{\ ?\ }$

    A. 16 oz   C. 50 oz

    B. 24 oz   D. 80 oz

12. $98.6°\text{F} = \underline{\ ?\ }$

    A. freezing point   C. body temperature

    B. boiling point   D. room temperature

Mark the equivalent expression.

13. $3 \times 4 = \underline{\ ?\ }$

   A. 4×4 C. $2 + 8$
   B. $6 + 6$ D. all of the above

14. $56 \times 8 = \underline{\ ?\ }$

   A. $56 \div 8$ C. $50 + (6 + 8)$
   B. $(50 \times 8) + (6 \times 8)$ D. all of the above

15. $63 \div 7 = \underline{\ ?\ }$

   A. 3×3 C. $5 + 4$
   B. $15 - 6$ D. all of the above

16. Jon ate 2 more cookies.

   A. $c + c$ C. $c + 2$
   B. $c \cdot 2$ D. all of the above

17. $30 + 8 = \underline{\ ?\ }$

   A. $20 + 18$ C. $20 + 28$
   B. $30 + 18$ D. all of the above

18. $25 - 10 = \underline{\ ?\ }$

   A. 3×5 C. $30 \div 2$
   B. $10 + 5$ D. all of the above

Mark the value for n.

19. $n + 10 = 30$

   A. 10   C. 30
   B. 20   D. none of the above

20. $n \times n = 25$

   A. 15   C. 5
   B. 10   D. none of the above

21.

12		
n	n	n

   A. 3   C. 6
   B. 4   D. none of the above

22.

n			
10	10	10	10

   A. 40   C. 20
   B. 30   D. none of the above

23.

100	
25	n

   A. 10   C. 50
   B. 25   D. none of the above

24. $9\overline{)72}^{\,n}$

   A. 8   C. 4
   B. 6   D. none of the above

25. $9 \times 1 = n$

   A. 0   C. 9
   B. 1   D. none of the above

Multiplying a Whole Number and a Fraction

Name _____

Multiplication is the same as repeated addition. Multiply the **number of addends** times the **numerator** of the fraction. Simplify.

$$6 \times \frac{3}{4} = \frac{6 \times 3}{4} = \frac{18}{4} = 4\frac{2}{4} = 4\frac{1}{2}$$

or

$$6 \times \frac{3}{4} = \frac{3}{4} + \frac{3}{4} + \frac{3}{4} + \frac{3}{4} + \frac{3}{4} + \frac{3}{4}$$

Solve by writing an addition equation. Write the answer in lowest terms.

1. $5 \times \frac{3}{12} =$

   $\frac{3}{12} + \frac{3}{12} + \frac{}{12} + \frac{}{12} + \frac{}{12} = \frac{}{12} =$

2. $2 \times \frac{1}{2} =$

   $\frac{}{} + \frac{}{} = \frac{}{}$

3. $3 \times \frac{3}{7} =$

4. $4 \times \frac{5}{9} =$

Shade the circles. Solve. Write the answer in lowest terms.

5. Shade $\frac{3}{8}$ of a circle. Shade another $\frac{3}{8}$. Continue until you have shaded $\frac{3}{8}$ four times.

   $4 \times \frac{3}{8} = \frac{12}{8} =$

6. Shade $\frac{5}{12}$ of a circle six times. Solve.

   $6 \times \frac{5}{12} =$

Solve. Write the answer in lowest terms.

7. $9 \times \frac{1}{3} = \frac{9 \times 1}{3} = \frac{9}{3} = 3$

8. $3 \times \frac{2}{5} =$

9. $5 \times \frac{4}{8} =$

10. $2 \times \frac{5}{12} =$

11. $5 \times \frac{2}{3} =$

12. $6 \times \frac{2}{4} =$

Solve. Write the answer in lowest terms. Label the answer.

13. If there are 5 people in your family, and you each have $\frac{1}{8}$ of a cake, how much cake does your family have altogether?

14. Marlene wants to cut back on sugar, so she decided to use $\frac{2}{3}$ of the amount of sugar that the recipe requires. The recipe calls for 2 cups of sugar. How much sugar will she need?

Use the chart to find the answer.

1. John and Sara Hutcheson are planning to take the youth group to Freddy's Fun Park for an all-day activity. How much will it cost for 33 young people to go for a single day?

Freddy's Fun Park	
Ticket	Price
Single day (ages 3–61)	$60
Single day (ages 62+)	$43
Season pass	$90

2. There will be 8 adult chaperones. Two of them are over 62. What is the total cost for the chaperones to go for a single day?

3. If the Hutchesons plan to take the young people on this trip 2 times this summer, would it be better to purchase 2 single-day tickets or 1 season pass per person? Why?

4. What would be less expensive for the chaperones over 62: 2 single-day tickets or 1 season pass? Why?

Write the measurement of the unknown angle.

5.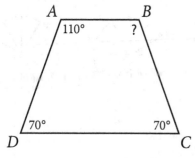

$$110° + \angle B + 70° + 70° = 360°$$

$$\angle B + \underline{\hspace{1cm}} = 360°$$

$$360° - \underline{\hspace{1cm}} = \underline{\hspace{1cm}}$$

$$\angle B = \underline{\hspace{1cm}}$$

6.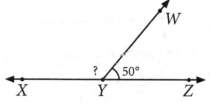

$\angle XYW = \underline{\ ?\ }$	$\angle WYZ = 50°$

$$\underline{\hspace{1cm}} + n = 180°$$

$$n = 180° - \underline{\hspace{1cm}}$$

$$n = \underline{\hspace{1cm}}$$

$$\angle XYW = \underline{\hspace{1cm}}$$

Rename to an equivalent fraction in lowest terms.

7. $\frac{2}{8} =$

8. $\frac{10}{20} =$

9. $\frac{8}{24} =$

10. $\frac{3}{9} =$

Rename as a mixed number in lowest terms.

11. $\frac{7}{2} =$

12. $\frac{17}{10} =$

13. $\frac{8}{3} =$

14. $\frac{10}{4} =$

Finding a Fraction of a Whole Number

Name _____

Solve and label your answer. Circle parts of the picture to show the solution.

1. Jason had a bag with 16 chocolate caramels. Jason told Ben that he could have $\frac{3}{4}$ of his caramels. How many caramels did Jason give Ben?

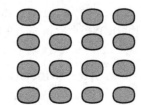

Solve. Write the answer in lowest terms. Label the answer. Draw a picture to show the solution.

2. | $\frac{1}{6}$ of a dozen rolls |

   $\frac{1}{6} \times 12 =$

3. | $\frac{3}{4}$ of 40 baseballs |

   $\frac{3}{4} \times 40 =$

4. | $\frac{2}{5}$ of 20 pencils |

   $\frac{2}{5} \times 20 =$

Solve. Write the answer in lowest terms.

5. $\frac{2}{3} \times 9 =$

6. $\frac{3}{7} \times 8 =$

7. $\frac{2}{6} \times 14 =$

8. $\frac{6}{10} \times 7 =$

9. $\frac{3}{8} \times 20 =$

10. $\frac{5}{9} \times 5 =$

Write an equation to find the answer. Solve. Label the answer.

11. How many minutes are in $\frac{1}{2}$ of an hour?

    $\frac{1}{2}$ of $60 = \frac{1 \times 60}{2} =$

12. How many months are in $\frac{2}{3}$ of a year?

    $\frac{2}{3}$ of $12 = \frac{2 \times 12}{3} =$

13. How many eggs are in $\frac{1}{3}$ of a dozen?

14. How many inches are in $\frac{3}{4}$ of a foot?

Solve by writing an addition equation.

15. $2 \times \frac{4}{5} =$

16. $7 \times \frac{3}{8} =$

Divisibility Rules

A number is divisible by . . .

2 if the number is an even number.
3 if the digits of the number add up to a multiple of 3.
4 if the last 2 digits in the number are divisible by 4.

5 if the number has a 5 or a 0 in the Ones place.
6 if the number is divisible by both 2 and 3.
10 if the number has a 0 in the Ones place.

Write the number or numbers of the divisibility rules that apply to the given number.

1. 54: _2, 3, 6_   2. 28: _____   3. 120: _____

4. 60: _____   5. 55: _____   6. 255: _____

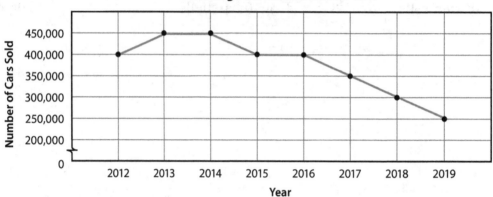

Average Car Sales for 2012–19

Use the line graph to answer the question.

7. About how many cars were sold in 2012?

_____

8. Which two years' sales stayed the same as the year before?

_____

9. Between which two years was there an increase of 50,000 cars sold?

_____

10. What was the number of cars sold in 2019?

_____

Write two different mathematical expressions for the number.

11. | 25 |

_____
_____

12. | 33 |

_____
_____

13. | 40 |

_____
_____

14. | 9 |

_____
_____

Write an equation. Solve. Label when applicable.

15. Each of 4 players scored 23 points in the game. Another player scored 15 points. How many points were scored in all?

16. Lori improved her history test score by 13 points from the last test. Her recent score was 102. What was her previous test score?

Finding a Fraction of a Fraction

Name _____

Solve. Write the answer in lowest terms. Match the diagram to the corresponding equation.

_____ 1. $\frac{1}{3} \times \frac{2}{3} =$ _____

_____ 2. $\frac{3}{5} \times \frac{1}{2} =$ _____

_____ 3. $\frac{4}{5} \times \frac{2}{3} =$ _____

_____ 4. $\frac{3}{4} \times \frac{4}{7} =$ _____

A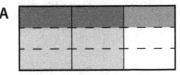

B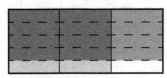

C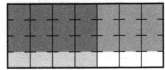

D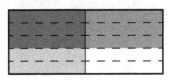

Solve by multiplying the numerators and multiplying the denominators. Write the answer in lowest terms.

5. $\frac{2}{5} \times \frac{1}{2} =$

6. $\frac{7}{10} \times \frac{1}{5} =$

7. $\frac{4}{5} \times \frac{1}{2} =$

8. $\frac{2}{3} \times \frac{1}{6} =$

9. $\frac{4}{8} \times \frac{1}{12} =$

10. $\frac{3}{4} \times \frac{2}{3} =$

Solve. Write the answer in lowest terms. Label the answer.

11. Vayle and his family went out to dinner. For dessert Vayle ate $\frac{1}{2}$ of his slice of cheesecake. A slice of cheesecake is $\frac{1}{16}$ of a whole cake. How much of the whole cheesecake did he eat?

12. The rest of the family shared 2 slices of cheesecake. How much of the whole cheesecake did they eat?

Solve. Write the answer in lowest terms.

13. $5 \times \frac{1}{7} =$

14. $\frac{4}{5} \times \frac{2}{7} =$

15. $15 \times \frac{1}{8} =$

16. $\frac{2}{5} \times 9 =$

17. $8 \times \frac{3}{7} =$

18. $\frac{2}{4} \times \frac{1}{8} =$

19. $\frac{3}{5} \times 12 =$

20. $\frac{3}{10} \times \frac{3}{4} =$

21. $4 \times \frac{3}{4} =$

Complete the table. Write the answer in lowest terms.

1.

Rule: $\times \frac{3}{4}$	
Input	Output
2	
4	
7	

2.

Rule: $\times \frac{1}{5}$	
Input	Output
3	
5	
10	

3.

Rule: $\times \frac{3}{8}$	
Input	Output
6	
8	
9	

Solve. Label your answer.

4. Thirty-nine classes at Calvary Christian School have collected 819 care packages to send to soldiers overseas. What was the average number of packages collected by each class?

5. Lennie scored a total of 2,052 points during four years of playing college basketball. He averaged 18 points a game. How many games did he play?

Use the circle graph to answer the question.

> The circle graph shows the types of pets owned by fifth graders at Grace Christian School.

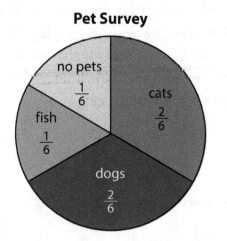

Pet Survey

6. What are the two most popular kinds of pets?

_____

7. What pet is owned by more students: dogs or fish?

_____

8. What fraction of the students do not have pets?

_____

Solve. Show your work.

9.

the sum of 12 and 15

10.

the sum of 15 and 27

11.

the difference between 83 and 16

12.

the difference between 69 and 19

Math 5 Activities

Multiplying a Mixed Number

Name _____

Solve by renaming the mixed number as an improper fraction.

1. $5 \times 1\frac{1}{2} =$

$5 \times \frac{3}{2} = \frac{15}{2} = 7\frac{1}{2}$

2. $3 \times 2\frac{3}{4} =$

3. $2 \times 1\frac{2}{3} =$

4. $2 \times 3\frac{2}{6} =$

5. $5 \times 1\frac{1}{4} =$

6. $2 \times 2\frac{4}{9} =$

Solve by renaming the mixed number as an improper fraction. Write the answer in lowest terms. Check the product by adding.

7. $2 \times 2\frac{1}{4} =$

$2 \times \frac{9}{4} =$

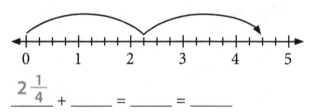

$2\frac{1}{4}$

_____ + _____ = _____ = _____

8. $3 \times 2\frac{1}{2} =$

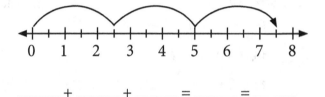

_____ + _____ + _____ = _____ = _____

Use the illustration to find the product. Label the answer.

9. Philip is building a treehouse. He cut a board into 3 pieces that were $2\frac{3}{4}$ feet long each. How long was the board before he cut it?

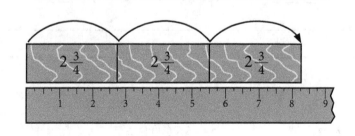

10. Mr. Watts made 4 deep-dish pizzas for a party. There were $1\frac{1}{2}$ pounds of cheese on each pizza. How much cheese did he use altogether?

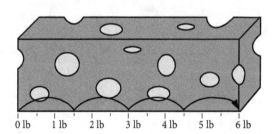

Solve. Write the answer in lowest terms.

11. $\frac{2}{3} \times \frac{5}{9} =$

12. $3 \times \frac{6}{7} =$

13. $\frac{1}{8} \times \frac{4}{6} =$

Solve. Write the answer in lowest terms.

1. $\frac{2}{7} \times 15 =$

2. $\frac{2}{6} \times 9 =$

3. $\frac{3}{4} \times 20 =$

4. $\frac{4}{9} \times 10 =$

5. $\frac{1}{2} \times 11 =$

6. $\frac{11}{12} \times 3 =$

7. $\frac{1}{3} \times 60 =$

8. $\frac{4}{5} \times 10 =$

9. $\frac{2}{3} \times 15 =$

10. $\frac{1}{8} \times 26 =$

11. $\frac{1}{6} \times 19 =$

12. $\frac{2}{9} \times 8 =$

Match the property to the equation.

_____ 13. $3 \cdot 8 = 8 \cdot 3$

_____ 14. $(6 \cdot 5) \cdot 4 = 6 \cdot (5 \cdot 4)$

_____ 15. $5 \cdot 1 = 5$

_____ 16. $42 \cdot 0 = 0$

_____ 17. $10 \cdot 0 = 0$

_____ 18. $9 \cdot 1 = 9$

_____ 19. $6 \cdot 4 = 4 \cdot 6$

_____ 20. $30 \cdot (2 \cdot 1) = (30 \cdot 2) \cdot 1$

Multiplication Properties

A Commutative Property

B Associative Property

C Zero Property

D Identity Property

Write the sequence for solving the equation. Solve.

Order of Operations
1. parentheses
2. multiplication or division
3. addition or subtraction

21. $30 - 3 \times 9 + 5 =$ _____

22. $(52 + 8) \times 2 =$ _____

23. $52 + (8 \times 2) =$ _____

24. $5 + 4 \times 3 \div 6 =$ _____

25. $39 - 3 \times (2 + 1) =$ _____

Write +, −, ×, or ÷ to make the statement true.

26. $7 \bigcirc 3 = 21$

27. $16 \bigcirc 8 = 2$

28. $56 \bigcirc 8 = 7$

29. $42 \bigcirc 10 = 32$

30. $16 \bigcirc 8 = 8$

31. $7 \bigcirc 7 = 49$

32. $48 \bigcirc 6 = 8$

33. $16 \bigcirc 8 = 24$

34. $7 \bigcirc 7 = 0$

35. $9 \bigcirc 9 = 18$

36. $16 \bigcirc 8 = 128$

37. $7 \bigcirc 7 = 14$

Multiplying Mixed Numbers

Name _____

Round each factor to the nearest whole number. Estimate the product.

1.

$3\frac{5}{6} \times 1\frac{1}{2} =$ ◯

2.

$4\frac{1}{5} \times 2\frac{1}{3} =$ ◯

3.

$1\frac{2}{3} \times 5\frac{1}{4} =$ ◯

4.

$6\frac{5}{6} \times 2\frac{3}{4} =$ ◯

5.

$7\frac{3}{4} \times 3\frac{1}{5} =$ ◯

6.

$2\frac{3}{5} \times 1\frac{6}{7} =$ ◯

Solve by renaming the mixed numbers. Write the answer in lowest terms.

Step 1
Rename mixed numbers as improper fractions.

Step 2
Multiply.

Step 3
Write the answer in lowest terms.

7. $1\frac{5}{6} \times 2\frac{1}{4} = \frac{?}{}$

$(\frac{1 \times 6}{6} + \frac{5}{6}) \times (\frac{2 \times 4}{4} + \frac{1}{4}) = \frac{?}{}$

$(\frac{6}{6} + \frac{5}{6}) \times (\frac{}{4} + \frac{}{4}) = \frac{?}{}$

$\frac{}{6} \times \frac{}{4} = \frac{}{24}$

$\frac{}{24} = 4\frac{}{24} = 4\frac{}{}$

8. $2\frac{1}{5} \times 2\frac{1}{2} = \frac{?}{}$

$(\frac{2 \times 5}{5} + \frac{}{}) \times (\frac{\times}{} + \frac{}{}) = \frac{?}{}$

$(\frac{}{} + \frac{}{}) \times (\frac{}{} + \frac{}{}) = \frac{?}{}$

$\frac{}{} \times \frac{}{} = \frac{}{}$

$\frac{}{} = \frac{}{} = \frac{}{}$

9. $1\frac{2}{3} \times 4\frac{1}{2} = \frac{?}{}$

$(\frac{\times}{} + \frac{}{}) \times (\frac{\times}{} + \frac{}{}) = \frac{?}{}$

$(\frac{}{} + \frac{}{}) \times (\frac{}{} + \frac{}{}) = \frac{?}{}$

$\frac{}{} \times \frac{}{} = \frac{}{}$

$\frac{}{} = \frac{}{} = \frac{}{}$

10. $2\frac{1}{3} \times 4\frac{1}{2} = \frac{?}{}$

11. $2\frac{2}{4} \times 1\frac{1}{5} = \frac{?}{}$

Write an equation for the sentence. Solve. Write the answer in lowest terms. Label the answer.

1. Jane ate $\frac{2}{3}$ of $\frac{1}{2}$ a banana. _____$\frac{2}{3} \times \frac{1}{2}$_____

2. Zoe scrambled $\frac{3}{4}$ of a dozen eggs. _____

3. Michael's car used $\frac{2}{3}$ of $\frac{3}{4}$ of a tank of gas. _____

4. Kelsey's students ate $\frac{3}{8}$ of 2 dozen cookies. _____

5. A sweater was on sale for $\frac{2}{5}$ of $\frac{1}{2}$ the price. _____

6. How many days are in $\frac{4}{7}$ of a week? _____

7. Jamal ate $\frac{1}{2}$ of $\frac{6}{8}$ of the pizza. _____

Match.

A acute   **B** obtuse   **C** right

8.
_____

9.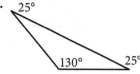
_____

10.
_____

11.
_____

12.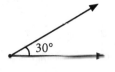
_____

13.
_____

14.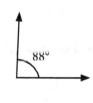
_____

15.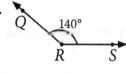
_____

Complete the fact.

16. 1 foot = _____ inches   17. 1 yard = _____ feet   18. 1 pound = _____ ounces

19. 1 mile = _____ yards   20. 1 yard = _____ inches   21. 1 gallon = _____ quarts

22. 1 mile = _____ feet   23. 1 ton = _____ pounds   24. 1 pint = _____ cups

25. 1 hour = _____ minutes   26. 1 year = _____ days   27. 1 quart = _____ pints

Solve. Label your answer.

28. Janice spent 432 minutes spring-cleaning her house. How many hours did she clean?

432 min =   hr   min

29. Wayne spent all day Saturday fencing the pasture. He built a total of 1,044 feet of fence. How many yards of fence did he build?

Dividing a Whole Number by a Fraction

Name _____

Use the picture to solve the division equation.

1. $3 \div \frac{1}{3} =$ _____

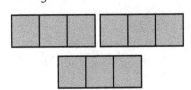

2. $2 \div \frac{2}{4} =$ _____

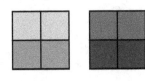

3. $1 \div \frac{2}{8} =$ _____

Use the picture to solve the division equation. Multiply to check the answer.

4. $5 \div \frac{1}{5} =$ _____

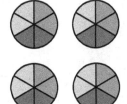

5. $4 \div \frac{2}{6} =$ _____

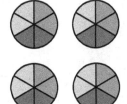

6. $6 \div \frac{3}{4} =$ _____

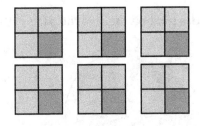

$$25 \times \frac{1}{5} = \frac{25}{5} =$$

Write an equation for the problem. Draw a picture to solve. Label your answer.

7. Mrs. Moran purchased a 5-pound bag of miniature chocolate candies. She gave each student a $\frac{1}{3}$ pound bag of the candy. How many students does she have?

8. Mr. Carruthers purchased 6 pies for the pie-eating contest after school. The pies were cut into sixths. How many slices were there in all?

Show the jumps on the number line to solve.

9. $4 \div \frac{2}{3} =$ _____

10. $3 \div \frac{1}{2} =$ _____

11. $3 \div \frac{3}{4} =$ _____

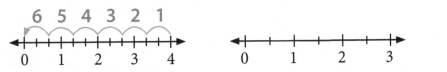

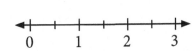

Rename the mixed number as an improper fraction. Solve.

1. $5 \times 1\frac{5}{7} =$

2. $3 \times 2\frac{1}{2} =$

3. $2 \times 1\frac{1}{3} =$

4. $2 \times 3\frac{1}{2} =$

5. $4 \times 1\frac{2}{3} =$

6. $3 \times 2\frac{2}{3} =$

7. What is the product of 3 and $2\frac{5}{6}$? _____

8. Five packages that each weigh $2\frac{3}{8}$ pounds equal a total of _____ pounds.

Use symbols to name the parts of the circle.

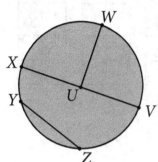

9. This is circle _____.

10. Name the 3 radii. _____, _____, _____

11. The diameter is _____.

12. Name a chord. _____

13. Name the two radii that make up the diameter. _____, _____

List the factors of each pair of numbers. List the factors they have in common.

14. 4: _____

8: _____

_____

15. 21: _____

6: _____

_____

16. 27: _____

81: _____

_____

Solve. Show your work.

17.
the quotient of 2,778 and 6

18.
the quotient of 50,496 and 8

19.
the product of 3,846 and 5

20.
the product of 1,002 and 7

Dividing a Fraction by a Fraction

Name _____

Use the picture to help you solve the problem.

1. $\frac{3}{6} \div \frac{1}{6} =$ _____

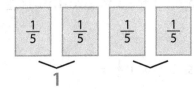

2. $\frac{4}{5} \div \frac{2}{5} =$ _____

3. $\frac{6}{8} \div \frac{3}{8} =$ _____

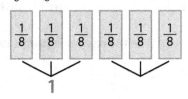

Draw a number line to model the problem. Solve.

4. $\frac{3}{8} \div \frac{1}{8} =$ __3__

5. $\frac{4}{6} \div \frac{2}{6} =$ _____

6. $\frac{2}{4} \div \frac{1}{8} =$ _____

Draw a picture to model the division equation. Solve.

7. $6 \div \frac{1}{2} =$ _____

8. $4 \div \frac{1}{4} =$ _____

9. $2 \div \frac{3}{6} =$ _____

Solve. Write the answer in lowest terms.

10. $6\frac{2}{3} \times \frac{5}{9} =$

11. $\frac{1}{9} \times 2\frac{1}{2} =$

12. $\frac{4}{5} \times \frac{1}{4} =$

13. $\frac{1}{2} \times 2 =$

14. $3\frac{6}{7} \times 1\frac{1}{8} =$

15. $3\frac{1}{3} \times 4\frac{1}{2} =$

Write the mathematical expression for the phrase.

1. the product of three and one-fourth and four _____

2. a one-and-one-fourth pound piece of a two-and-one-half pound block of cheese _____

3. four and one-eighth multiplied by three and two-thirds _____

Solve. Write the answer in lowest terms.

4. $6\frac{2}{3} \times 4\frac{1}{2} =$

5. $3\frac{1}{3} \times 1\frac{1}{4} =$

6. $2\frac{1}{5} \times 5\frac{2}{5} =$

Plot and label the points on the graph. Draw the figure.

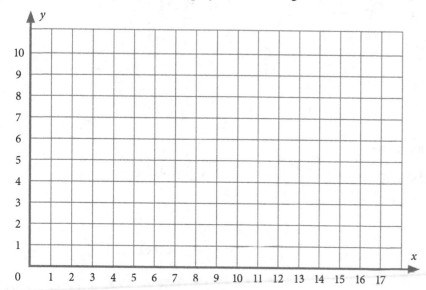

7. A (1, 3)

8. B (1, 9)

9. C (3, 2)

10. D (3, 10)

11. E (6, 3)

12. F (6, 7)

13. G (9, 4)

14. H (9, 6)

15. Draw $\overleftrightarrow{AB}$.

16. Draw $\overleftrightarrow{CD}$.

17. Draw $\overleftrightarrow{EH}$.

18. Draw $\overleftrightarrow{FG}$.

Plot and label the points on the graph above.

19. I (16, 10)

20. J (13, 8)

21. K (15, 4)

22. L (13, 2)

Solve. Label your answer.

23. Brandon and Liz just built a new addition to their house. It measures 20 feet by 14 feet. They want to put ceramic tile on the floor. How many square feet do they need?

24. Each tile costs $3.59 per square foot. How much will the tile cost?

25. Their budget for the floor is $1,000. Do they have enough money for the floor? Why or why not?

26. Brandon also had crown molding installed all around the room. What is the perimeter of the room?

Using Reciprocals to Divide Fractions

Name _____

Two numbers are **reciprocals** if their product equals 1.

$$\frac{2}{3} \times \frac{3}{2} = \frac{2 \times 3}{3 \times 2} = \frac{6}{6} = 1$$

The reciprocal of a fraction (not equal to 0) is found by interchanging the numerator and denominator.

Write the reciprocal.

1. $\frac{6}{8}$ $\boxed{\frac{8}{6}}$

2. $\frac{4}{9}$ $\boxed{\frac{9}{4}}$

3. $\frac{6}{8}$ ▢

4. $\frac{5}{7}$ ▢

5. $\frac{3}{4}$ ▢

6. $\frac{1}{2}$ ▢

7. $\frac{7}{9}$ ▢

8. $\frac{8}{12}$ ▢

9. $\frac{9}{10}$ ▢

10. $\frac{11}{14}$ ▢

Write the reciprocal to complete the multiplication equation.

11. $\frac{2}{3} \times$ ▢ $= 1$

12. $\frac{5}{8} \times$ ▢ $= 1$

13. $\frac{10}{16} \times$ ▢ $= 1$

14. $\frac{1}{8} \times$ ▢ $= 1$

Dividing fractions is the same as multiplying by the reciprocal of the divisor.

Rewrite the equation to multiply by the reciprocal. Solve. Write the answer in lowest terms.

15. $\frac{2}{8} \div \frac{1}{8} = \frac{2}{8} \times \frac{8}{1} = \frac{16}{8} = 2$

16. $3 \div \frac{3}{5} =$

17. $\frac{3}{9} \div \frac{8}{9} =$

18. $\frac{2}{7} \div \frac{4}{5} =$

19. $3 \div \frac{3}{8} =$

20. $10 \div \frac{4}{5} =$

Solve. Write the answer in lowest terms. Label your answer.

21. Sam had 3 candy bars. He cut them into $\frac{1}{3}$ pieces. How many pieces did he have altogether?

22. Bonnie is working on a craft project. She has 12 feet of yarn that needs to be cut into $\frac{1}{4}$ foot pieces. How many pieces will she have?

Complete the table.

1.

Rule: × $\frac{1}{3}$	
4	
6	
8	
10	

2.

Rule: × $\frac{1}{5}$	
3	
5	
7	
9	

3.

Rule: × $\frac{1}{2}$	
2	
4	
7	
11	

Solve. Show your work.

- Line up decimal points.
- Annex 0s if needed.
- Place a decimal point in the answer.

4. the sum of 7.809 and 4.002

5. the difference between 4.22 and 1.99

6. the difference between 15 and 9.86

7. the sum of 1.32 and 3.147

8. the difference between 5.32 and 4.2

9. the sum of 32, 7.07, and 42.63

10. the sum of $81.19 and $9.40

11. the quotient of 4,390 and 22

12. Use multiplication to check problem 11.

13. the quotient of 1,358 and 9

14. Use multiplication to check problem 13.

Write +, −, ×, or ÷ to complete the equation.

15. 12 ◯ 8 = 20

16. 10 ◯ 12 = 120

17. 72 ◯ 9 = 8

18. 29 ◯ 2 = 27

Dividing Fractions

Name _____

Dividing a Fraction by a Fraction

1. Find the reciprocal by inverting the fraction.
2. Multiply by the reciprocal.
3. Simplify.
4. Check.

Multiply by the reciprocal to find the quotient. Write the answer in lowest terms.

1. $\frac{2}{3} \div \frac{1}{3} = \frac{2}{3} \times \frac{3}{1} = \frac{6}{3} = 2$

2. $\frac{5}{7} \div \frac{4}{5} = $ ___ $\times$ ___ $=$

3. $\frac{1}{2} \div 5 =$

   ___ $\times$ ___ $=$

4. $\frac{2}{5} \div \frac{3}{4} =$

   ___ $\times$ ___ $=$

5. $\frac{1}{2} \div \frac{1}{3} =$

   ___ $\times$ ___ $=$

6. $5 \div \frac{2}{3} =$

7. $2\frac{3}{4} \div 1\frac{2}{4} =$

8. $\frac{2}{3} \div 10 =$

Solve. Write the answer in lowest terms. Label the answer.

9. Mr. Peterson spends $1\frac{1}{4}$ hours each day taping trim before he starts to paint. How many hours does he spend in five days?

10. Mr. Peterson spends 4 hours each week cleaning his equipment. If he cleans for $1\frac{1}{3}$ hours each day, how many days does he clean each week?

Write an equation for the phrase. Solve. Write the answer in lowest terms.

11. one-third of an hour in minutes _____

12. six feet of ribbon cut into one-third foot pieces _____

13. the product of one-eighth and three-fourths _____

14. the difference of six-eighths and two-fourths _____

15. six sets of one-fifth _____

Use the picture to help you solve the problem.

1. $\frac{4}{6} \div \frac{2}{6} =$ _____

2. $\frac{3}{5} \div \frac{1}{5} =$ _____

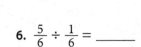

3. $\frac{1}{2} \div \frac{2}{4} =$ _____

4. $\frac{6}{8} \div \frac{2}{8} =$ _____

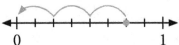

5. $\frac{2}{3} \div \frac{1}{3} =$ _____

6. $\frac{5}{6} \div \frac{1}{6} =$ _____

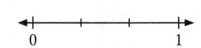

Use the pictograph to answer the question.

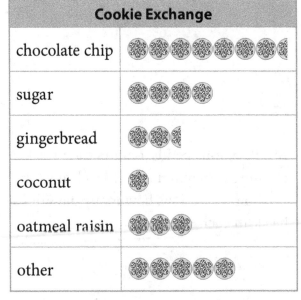

Cookie Exchange — chocolate chip, sugar, gingerbread, coconut, oatmeal raisin, other

= 2 dozen cookies

The women's ministry sponsored a cookie exchange.

7. How many dozen chocolate chip cookies were brought?

_____

8. How many more dozen sugar cookies were there than oatmeal raisin cookies?

_____

9. How many dozen cookies were there altogether?

_____

10. How many dozen sugar cookies and gingerbread cookies were there altogether?

_____

Write the elapsed time or the time that is described.

11. Jake started mowing the lawn at 9:00 a.m. and finished at 1:15 p.m. _____

12. School starts at 8:00 a.m. and ends at 2:45 p.m. _____

13. Elise had job interviews scheduled from 11:59 a.m. to 5:05 p.m. _____

14. Peggy left work 7 hours and 45 minutes after she arrived at 8:00 a.m. _____

15. Church services ended 1 hour and 23 minutes after 6:45 p.m. _____

16. Raphael arrived in Nova Scotia 5 hours and 23 minutes after leaving the airport at 6:22 a.m. _____

17. Laurie went to bed 14 hours and 15 minutes after she woke up that morning at 7:05 a.m. _____

Fractions: Multiplication & Division Review

Name _____

Mark the answer.

1. $12 \times \frac{1}{6} = \underline{\ ?\ }$

   ○ 6   ○ 2

   ○ $12\frac{1}{6}$ ○ 3

2. $3 \times \frac{5}{9} = \underline{\ ?\ }$

   ○ $1\frac{2}{3}$ ○ $\frac{8}{9}$

   ○ $\frac{5}{27}$ ○ $1\frac{1}{3}$

3. $7 \times \frac{1}{4} = \underline{\ ?\ }$

   ○ $\frac{4}{7}$ ○ $1\frac{1}{2}$

   ○ $1\frac{3}{4}$ ○ $1\frac{1}{7}$

Mark the fraction that is in lowest terms.

4. ○ $4\frac{2}{4}$ ○ $4\frac{15}{20}$

   ○ $4\frac{1}{2}$ ○ $4\frac{6}{8}$

5. ○ $1\frac{3}{9}$ ○ $5\frac{2}{4}$

   ○ $8\frac{4}{6}$ ○ $3\frac{1}{3}$

6. ○ $\frac{14}{4}$ ○ $1\frac{9}{12}$

   ○ $4\frac{6}{8}$ ○ $\frac{5}{6}$

Solve. Write the answer in lowest terms. Label the answer.

7. Mrs. Rosier is baking cookies for open house at school. Each recipe calls for $\frac{3}{4}$ of a cup of brown sugar. How much brown sugar does she need to make 3 recipes?

Mark the answer.

8. $2\frac{5}{7} \times \frac{1}{5} = \underline{\ ?\ }$

   ○ $\frac{19}{35}$ ○ $\frac{14}{35}$

   ○ $2\frac{5}{35}$ ○ $1\frac{18}{35}$

9. $\frac{1}{2} \times 2\frac{4}{9} = \underline{\ ?\ }$

   ○ $4\frac{4}{9}$ ○ $1\frac{2}{9}$

   ○ $1\frac{1}{2}$ ○ $2\frac{5}{9}$

10. $\frac{1}{8} \times 3 = \underline{\ ?\ }$

    ○ $\frac{1}{24}$ ○ $\frac{3}{8}$

    ○ $\frac{4}{8}$ ○ $1\frac{5}{8}$

11. $\frac{2}{3} \times \frac{1}{4} = \underline{\ ?\ }$

    ○ $\frac{8}{3}$ ○ $\frac{2}{4}$

    ○ $1\frac{2}{3}$ ○ $\frac{1}{6}$

Mark the expression that matches the picture.

12.

    ○ $\frac{2}{5} \times 15$ ○ 3×5

    ○ $\frac{5}{15} \times 2$ ○ $\frac{2}{3} \times 15$

13.

    ○ $\frac{1}{3} \times 12$ ○ 3×12

    ○ $\frac{1}{4} \times 12$ ○ $3 \times \frac{1}{4}$

Mark the expression that matches the picture.

14.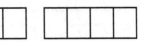

   ○ $2 \div \frac{1}{4}$ ○ $8 \div 1$

   ○ $4 \div 2$ ○ $8 \div 2$

15.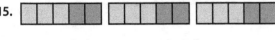

   ○ $3 \div \frac{1}{5}$ ○ $3 \div \frac{3}{5}$

   ○ $5 \div \frac{1}{3}$ ○ $5 \div 3$

16.

   ○ $6 \div \frac{2}{6}$

   ○ $1 \div \frac{2}{6}$

   ○ $1 \div \frac{1}{3}$

   ○ $6 \div 1$

17.

   ○ $3 \div \frac{4}{4}$ ○ $3 \div \frac{1}{4}$

   ○ $4 \div 3$ ○ 4×3

Solve. Label your answer.

18. Bonnie baked one-third the amount of cookies that Jennie baked. If Jennie baked 48 cookies, how many cookies did Bonnie bake?

19. There were one hundred students sitting in the auditorium. One-fourth of them were sitting in the first two rows. How many students were in the first two rows?

Complete the table.

20.

Rule: $\times \frac{1}{2}$	
Input	Output
6	
8	
12	
20	

21.

Rule: $\times \frac{2}{3}$	
Input	Output
3	
6	
9	
12	

22.

Rule: $\div \frac{1}{3}$	
Input	Output
3	
6	
9	
12	

23.

Rule: $\div \frac{1}{2}$	
Input	Output
2	
4	
6	
8	

Mark the answer.

1. Which list is the complete list of factors for 24?

   A. 1, 3, 4, 12

   B. 1, 2, 4, 8

   C. 1, 2, 3, 4, 6, 8, 12, 24

   D. none of the above

2. Which number is *not* a multiple of 4?

   A. 64

   B. 72

   C. 90

   D. 108

3. Which statement is true of the number 27?

   A. It is a composite number.

   B. It is divisible by 3.

   C. It is an odd number.

   D. all of the above

4. Which problem shows the Identity Property of Addition?

   A. $15.38 + p = 25.38$

   B. $15.38 + p = 15.38$

   C. $15.38 + p = 16.38$

   D. $15.38 + p = 15.48$

Mark the value of n.

5.

28			
n	n	n	n

   A. 4   C. 7

   B. 6   D. none of the above

6.

1,500	
n	n

   A. 525   C. 8.75

   B. 750   D. 1,000

7.

$\frac{1}{2}$					
n	n	n	n	n	n

   A. $\frac{1}{8}$ C. $\frac{1}{10}$

   B. $\frac{1}{9}$ D. $\frac{1}{12}$

8.

n			
1.25	1.25	1.25	1.25

   A. 4.75   C. 5.25

   B. 5   D. 5.50

9.

n		
$2\frac{1}{2}$	$2\frac{1}{2}$	$2\frac{1}{2}$

   A. 5   C. $8\frac{1}{2}$

   B. $7\frac{1}{2}$ D. 10

Mark the answer.

10.

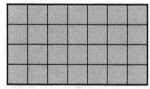

The perimeter is __?__.

A. 22 units   C. 30 units

B. 26 units   D. 34 units

11.

6 ft

The area is __?__.

A. 30 ft²   C. 42 ft²

B. 36 ft²   D. 48 ft²

12.

If the area of the rectangle is 400 in.², what is the area of the gray triangle?

A. 100 in.²   C. 250 in.²

B. 200 in.²   D. 300 in.²

13.

10 mm

20 mm

What shapes are created when a vertical line of symmetry is made?

A. squares   C. rectangles

B. triangles   D. a square and a rectangle

14.

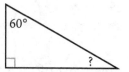

What is the measure of the unknown angle?

A. 30°   C. 50°

B. 40°   D. 60°

15.

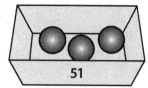

A. 13   C. 16

B. 15   D. 17

16.

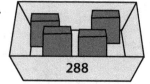

A. 50   C. 70

B. 68   D. 72

17. Scarlett is 2 years older than Kayla but 4 years younger than Hunter. How old is Kayla if Hunter is 17?

A. 13   C. 11

B. 12   D. 10

18. How many feet of ribbon does Monica have if she has 168 inches of ribbon?

A. 8 ft   C. 29 ft

B. 14 ft   D. 56 ft

19. Daniel drew a quadrilateral with only 2 right angles. Which figure could he have drawn?

A. rhombus   C. trapezoid

B. parallelogram   D. rectangle

Decimals

Place Value Chart					
Hundreds	Tens	Ones ·	Tenths	Hundredths	Thousandths
		3 ·	8	7	2

decimal form: 3.872

fraction form: $3\frac{872}{1,000}$

Write the decimal form of the given fraction.

1. Jeremy ate $\frac{5}{10}$ of the cookies. _____

2. Susan saved $\frac{49}{100}$ of her salary. _____

3. Vern drove $\frac{359}{1,000}$ of the miles on his trip. _____

4. Joyce worked $\frac{1}{2}$ of the day. _____

5. 633 years is $\frac{633}{1,000}$ of a millennium. _____

Write the fraction in decimal form.

6. $\frac{13}{100} = $ _____   7. $\frac{6}{10} = $ _____   8. $\frac{349}{1,000} = $ _____   9. $\frac{8}{10} = $ _____   10. $\frac{4}{100} = $ _____

11. $\frac{23}{1,000} = $ _____   12. $\frac{41}{1,000} = $ _____   13. $\frac{5}{100} = $ _____   14. $\frac{21}{100} = $ _____   15. $\frac{1}{10} = $ _____

Fill in the blank with the correct place value for the number.

25.304

16. What digit is in the Tens place? _____

17. What digit is in the Thousandths place? _____

18. What digit is in the Tenths place? _____

19. What digit is in the Ones place? _____

20. What digit is in the Hundredths place? _____

Write the decimal in fraction form.

21. 0.04 =

22. 0.15 =

23. 0.2 =

24. 0.432 =

25. 0.003 =

26. 0.1 =

Use a ruler to draw the figure. Use the symbol to name the figure.

1.
line

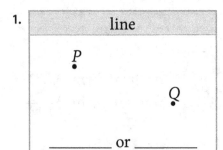

_____ or _____

2.
line segment

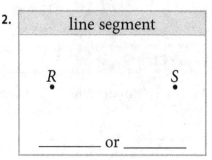

_____ or _____

3.
plane

plane _____

4.
ray

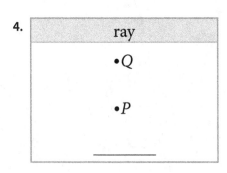

_____

5.
ray

_____

6.
angle

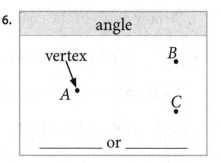

_____ or _____

Write +, −, ×, or ÷ to make the statement true.

7. $8 \bigcirc 4 = 32$

8. $63 \bigcirc 9 = 7$

9. $8 \bigcirc 8 = 16$

10. $62 \bigcirc 20 = 42$

11. $8 \bigcirc 5 = 40$

12. $4 \times 4 = 8 \bigcirc 2$

13. $5 \bigcirc 4 = 10 + 10$

14. $48 \bigcirc 8 = 12 \div 2$

15. $5 \bigcirc 5 = 5 \times 2$

Compare the figures. Write the symbol for similar (∼) or congruent (≅).

16.

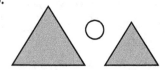

17.

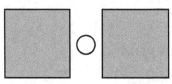

18.

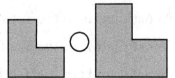

Complete the table.

19.
Rule: × 8	
Input	Output
5	
10	
20	
40	

20.
Rule: ÷ 7	
Input	Output
28	
42	
56	
70	

21.
Rule: − 6	
Input	Output
12	
20	
30	
42	

22.
Rule: + 9	
Input	Output
1	
8	
15	
22	

Rounding Decimals

Name _____

Write the equivalent fraction in hundredths. Write the equivalent decimal in hundredths.

1. $\frac{7}{10} = \boxed{\frac{70}{100}} = \underline{0.70}$

2. $\frac{3}{10} = \boxed{\frac{}{100}} = $ _____

3. $\frac{6}{10} = \boxed{\frac{}{100}} = $ _____

4. $\frac{8}{10} = \boxed{} = $ _____

5. $\frac{2}{10} = \boxed{} = $ _____

6. $\frac{4}{10} = \boxed{} = $ _____

Mark the location of the decimal on the number line with a dot. Round the decimal to the nearest hundredth.

7. 6.732 rounds to _____.

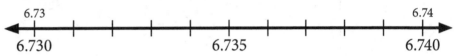

6.73   6.74
6.730   6.735   6.740

8. 4.938 rounds to _____.

4.93   4.94
4.930   4.935   4.940

Round the decimal to the nearest tenth.

9. 4.056 = _____

10. 5.89 = _____

11. 3.009 = _____

12. 2.761 = _____

Round the decimal to the nearest whole number.

13. 4.891 = _____

14. 6.321 = _____

15. 2.005 = _____

16. 9.099 = _____

Write the number in either decimal form or word form.

17. 5.14 = _____

18. one hundred seventy-four thousandths = _____

19. 8.34 = _____

20. nine and fifteen hundredths = _____

Solve.

21. Mrs. Garcia purchased $2\frac{1}{2}$ yards of blue material and 2.05 yards of red material. Which material did she buy more of? Explain.

_____

Use the symbol and 3 points to name the angle. Use a protractor to find the measurement of the angle. Classify the angle as *acute, obtuse, right,* or *straight.*

1.

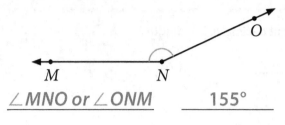

<u>∠MNO or ∠ONM</u>   <u>155°</u>

_____

2.

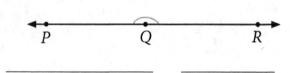

_____   _____

_____

3.

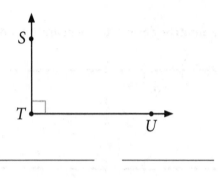

_____   _____

_____

4.

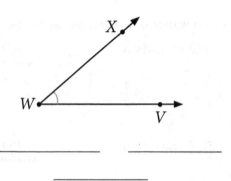

_____   _____

_____

Match.

A translation   **B** rotation   **C** reflection

5.

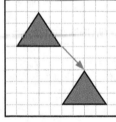

_____

6.

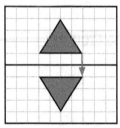

_____

7.

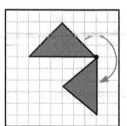

_____

8.

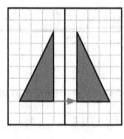

_____

Solve. Show your work.

9. What is the sum of 5.398 and 6.201?

10. What is the difference between 7.1 and 5.46?

11. What is the sum of 38.472, 5.9, and 0.13?

Comparing & Multiplying

Name _____

Solve. Label your answer.

1. Jesse cut 4 lengths of rope to use in a project. They measured 7.6 inches, 7.46 inches, 7.59 inches, and 7.64 inches. Write the lengths in order from shortest to longest.

   _____   _____   _____   _____

2. Mrs. Chin is making a quilt. She bought blue, red, white, and yellow fabric. The pieces measured 1.113 yards, 1.23 yards, 1.122 yards, and 1.132 yards. Order the measurements from least to greatest.

   _____   _____   _____   _____

Write >, <, or = to compare.

			Annex 0s to compare like decimals.
3. 2.135 ◯ 21.35	4. 6.808 ◯ 6.088	5. 4.090 ◯ 4.09	
6. 62.13 ◯ 6.213	7. 3.02 ◯ 3.020	8. 7.08 ◯ 708.1	

Estimate the product. Solve. Mark the correct answer.

9. ◯
$$\begin{array}{r} 3.8 \\ \times\ \ 4 \\ \hline ? \end{array}$$
◯ 1.52
◯ 15.2

10. ◯
$$\begin{array}{r} 4.12 \\ \times\ \ 5 \\ \hline ? \end{array}$$
◯ 206.0
◯ 20.60

11. ◯
$$\begin{array}{r} 8.74 \\ \times\ \ 3 \\ \hline ? \end{array}$$
◯ 26.22
◯ 262.2

12. ◯
$$\begin{array}{r} 1.068 \\ \times\ \ 4 \\ \hline ? \end{array}$$
◯ 4.272
◯ 42.72

13. ◯
$$\begin{array}{r} 6.5 \\ \times\ \ 3 \\ \hline ? \end{array}$$
◯ 19.5
◯ 1.95

14. ◯
$$\begin{array}{r} 4.10 \\ \times\ \ 6 \\ \hline ? \end{array}$$
◯ 2.460
◯ 24.60

Match.

_____ 15. 4.65

_____ 16. 0.465

_____ 17. 0.456

_____ 18. 4.605

_____ 19. 4.056

A $(4 \times 0.1) + (5 \times 0.01) + (6 \times 0.001)$

B $4\frac{56}{1,000}$

C 4 ones, 65 hundredths

D four hundred sixty-five thousandths

E four and six hundred five thousandths

Write the fraction form and decimal form.

1. Mrs. Todd has 4 sticks of gum left from a package of 10. _____ _____

2. Of the first 100 days of school, 50 days are already over. _____ _____

3. Javier has used 459 sheets from a package of 1,000. _____ _____

4. Uncle Alvin has lost 35 pounds of his goal of 100. _____ _____

5. Dennae has a goal of collecting $1,000 for charity.
   She has received $98 so far. _____ _____

Solve. Check by multiplying.

6.
625 pounds of sugar in 5-pound bags

$5\overline{)625}$ **Check**

_____ bags of sugar
_____ pounds remaining

7.
328 cookies in packs of a dozen

$12\overline{)328}$ **Check**

_____ packages of cookies
_____ cookies remaining

8.
710 students in 39 classrooms

$39\overline{)710}$ **Check**

_____ students in a classroom
_____ students remaining

Classify the triangle as *acute*, *obtuse*, or *right*.

9. a yield sign that has angles measuring 60°, 60°, and 60° _____

10. a kite with angles of 45°, 45°, and 90° _____

11. a head scarf with angles of 120°, 30°, and 30° _____

Solve.

12. $\frac{24}{8}$ = ____

13. 17 − 9 = ____

14. 8 + 9 = ____

15. 11 × 7 = ____

16. 12 × 3 = ____

17. 5 + 9 = ____

18. 36 ÷ 12 = ____

19. 16 − 9 = ____

20. sum of
    6 and 9 = ____

21. difference between
    19 and 8 = ____

22. product of
    6 and 9 = ____

23. quotient of
    56 and 8 = ____

Multiplying Decimals

Name _____

Use two colors to shade the parts to show the multiplication. Write the decimal equation in fraction form. Write the product in decimal form.

1.

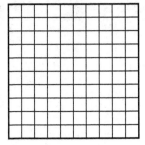

$0.5 \times 0.6 = \underline{\;?\;}$

$\dfrac{5}{10} \times \dfrac{6}{10} = \dfrac{}{100}$

product = _____

2.

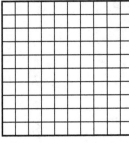

$0.2 \times 0.9 = \underline{\;?\;}$

$\dfrac{}{10} \times \dfrac{}{10} = \dfrac{}{}$

product = _____

3.

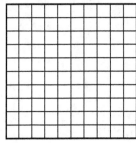

$0.6 \times 0.3 = \underline{\;?\;}$

$\dfrac{}{10} \times \dfrac{}{10} = \dfrac{}{}$

product = _____

Multiply. Count the decimals to find the decimal placement.

4. $\begin{array}{r} 0.4 \\ \times\, 0.7 \\ \hline \end{array}$

5. $\begin{array}{r} 1.3 \\ \times\, 1.4 \\ \hline \end{array}$

6. $\begin{array}{r} 4.29 \\ \times\; 1.3 \\ \hline \end{array}$

7. $\begin{array}{r} 19.8 \\ \times\; 5.1 \\ \hline \end{array}$

8. $\begin{array}{r} 0.21 \\ \times\; 4.3 \\ \hline \end{array}$

9. What is the product of 6.3 and 2.4? _____

10. What is the product of 0.25 and 4.2? _____

Circle the digit in the given place.

11. Hundredths place $3\,4\,.\,5\,7$

12. Ones place $1\,.\,7\,3\,2$

13. Tenths place $1\,4\,.\,8\,1\,5$

14. Thousandths place $0\,.\,8\,1\,2$

Write the value of 4 in the number.

15. 4.260 _____

16. 0.478 _____

17. 1.034 _____

18. 0.040 _____

Write the decimal in standard form.

19. three hundred sixty-four thousandths _____

20. two and fifty-three hundredths _____

Solve. Label your answer.

1. Sandy runs 8.25 miles every week. What is the total number of miles she will run in 9 weeks?

 Solve

2. Melanie jogs 9.5 laps around the track every day. She runs 5 days a week. How many laps does she run in one week?

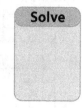 Solve

3. 6.398
   ×   5
   _____

4. 5.01
   ×   4
   _____

5. 7.8
   ×   9
   _____

6. 3.201
   ×   2
   _____

Draw a picture for the statement. Answer the question.

> The fifth-grade class had a pizza party for the last day of school. At the end of the party, the following amounts were left.

7. $2\frac{1}{8}$ of the mushroom pizzas were left.

8. $\frac{6}{8}$ of the sausage pizza was left.

9. $\frac{9}{8}$ of the pepperoni pizza was left.

10. How much sausage and pepperoni pizza was left in all? _____

11. Was there more mushroom pizza or pepperoni pizza left? _____

12. Which had more left: mushroom pizza or sausage and pepperoni pizzas combined?

   _____

Identify the triangles as *isosceles, scalene,* or *equilateral*.

13.
   5 ft
   _____

14.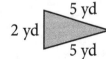
   5 yd
   2 yd
   5 yd
   _____

15.
   2 cm 4 cm
   3 cm
   _____

16.
   15 in.
   _____

Estimating & Multiplying

Name _____

Solve.

1. What is the standard form of 3 + 0.4 + 0.06 + 0.007? _____

2. What is the decimal form of $5\frac{186}{1,000}$? _____

3. Estimate the product of 1.7 and 3.1. _____

4. What is the product of 2.54 and 2.3? _____

Estimate the product. Solve.

5.
$$\begin{array}{r} 3.1 \\ \times\,5.7 \\ \hline \end{array}$$

6.
$$\begin{array}{r} 6.07 \\ \times\,\ 1.8 \\ \hline \end{array}$$

7.
$$\begin{array}{r} 3.1 \\ \times\,4.7 \\ \hline \end{array}$$

8.
$$\begin{array}{r} 1.8 \\ \times\,1.7 \\ \hline \end{array}$$

9.
$$\begin{array}{r} 3.29 \\ \times\,\ 2.7 \\ \hline \end{array}$$

Solve. Label your answer.

10. Teresa uses 3.5 ounces of deli meat for each sandwich. How many ounces would she need for 15 sandwiches?

11. Hardwood flooring costs $5 per square foot. How much would 14.5 square feet cost?

Use the chart to answer the question.

Gabrielle's Weight Loss	
week one	2.24 lb
week two	2.42 lb
week three	2.22 lb
week four	2.402 lb

12. During which week did Gabrielle lose the most weight?

_____

13. During which week did she lose the least? _____

14. During which week did she lose less: week two or week four?

_____

15. What is her total weight loss for all four weeks? _____

Write each decimal in fraction form. Multiply. Write the product in decimal form.

1. $0.6 \times 0.3 =$ __?__

$$\frac{6}{10} \times \frac{3}{10} = \frac{18}{100}$$

product = __0.18__

2. $0.9 \times 0.2 =$ __?__

$$\frac{}{10} \times \frac{}{10} = \frac{}{}$$

product = _____

3. $0.5 \times 0.4 =$ __?__

$$\frac{}{10} \times \frac{}{10} = \frac{}{}$$

product = _____

Multiply. Count the decimals to find the decimal placement.

4. $\begin{array}{r} 1.6 \\ \times\, 0.3 \\ \hline \end{array}$

5. $\begin{array}{r} 3.5 \\ \times\, 4.2 \\ \hline \end{array}$

6. $\begin{array}{r} 7.9 \\ \times\, 0.4 \\ \hline \end{array}$

7. $\begin{array}{r} 0.93 \\ \times\,\ 4.1 \\ \hline \end{array}$

8. $\begin{array}{r} 33.2 \\ \times\,\ 5.1 \\ \hline \end{array}$

Find the greatest common factor of the set of numbers.

9. List the factors.

16: _____

18: _____

The GCF of 16 and 18 is _____.

10. List the factors.

24: _____

36: _____

The GCF of 24 and 36 is _____.

11. Complete the factor tree.

15   25

15: _____

25: _____

The GCF of 15 and 25 is _____.

12. Complete the factor tree.

28   36

28: _____

36: _____

Multiply the common factors. _____

The GCF of 28 and 36 is _____.

Solve. Write the answer in lowest terms.

13. $3 \times \frac{1}{4} =$

14. $\frac{1}{6} \times \frac{2}{5} =$

15. $2 \times 1\frac{2}{5} =$

16. $\frac{2}{7} \times \frac{4}{5} =$

17. $\frac{3}{6} \times 4 =$

18. $\frac{4}{8} \times \frac{3}{4} =$

Dividing a Decimal by a 1-Digit Divisor

Name _____

Solve.

1. $5\overline{)31.5}$

2. $7\overline{)57.4}$

3. $3\overline{)9.69}$

4. $8\overline{)27.52}$

> Write the decimal point in the quotient.

Solve. Label your answer.

5. Josh and Rory ran laps for soccer training. Josh ran 7.7 laps on Monday, 5.8 laps on Wednesday, and 8.1 laps on Friday. What is his total number of laps?

_____

What was the average number of laps Josh ran each day for the week?

_____

6. Rory paced his laps during the week and ran 7.4 laps on Monday, Wednesday, and Friday. What is Rory's total number of laps?

_____

Who had the greater number of laps, Josh or Rory?

_____

Solve.

7.

Rule: ÷ 2	
Input	Output
13.2	
26.4	
32.8	
6.2	

8.

Rule: × 4	
Input	Output
1.42	
2.3	
4.01	
5.25	

9.

Rule: ÷ 5	
Input	Output
21.05	
33.90	
4.5	
1.25	

10.

Rule: × 6	
Input	Output
0.02	
0.25	
1.98	
6.21	

Round the decimal to the nearest tenth.

11. $4.006 =$ _____

12. $1.015 =$ _____

13. $0.063 =$ _____

14. $5.331 =$ _____

Round the decimal to the nearest hundredth.

15. $0.749 =$ _____

16. $0.362 =$ _____

17. $2.538 =$ _____

18. $3.893 =$ _____

19. Explain the difference between 0.5, 0.05, and 0.005.

_____

_____

_____

Estimate the product. Solve.

1.
   3.7
   × 5.2

2.
   2.6
   × 1.4

3.
   4.6
   × 2.9

4.
   5.1
   × 5.7

5.
   6.07
   × 3.8

6.
   3.1
   × 4.7

7.
   1.6
   × 1.4

8.
   2.7
   × 4

9.
   0.23
   × 0.8

10.
    1.07
    × 1.6

Solve. Show your work.

11. 3,900 soldiers were lined up in 12 rows.

12. 10,626 baseball cards were collected by 42 boys.

13. 15 students memorized a total of 3,645 verses.

There were an average of _____ soldiers in each row.

Each boy collected an average of _____ baseball cards.

Each student memorized an average of _____ verses.

Quotients Less than One

Divide to find the equivalent decimal for the fraction.

1. $\frac{1}{4} =$ _____

2. $\frac{1}{8} =$ _____

3. $\frac{2}{5} =$ _____

> Annex 0s as needed.

$$4\overline{)1.00}$$

4. $\frac{3}{6} =$ _____

5. $\frac{5}{8} =$ _____

6. $\frac{3}{8} =$ _____

7. $\frac{3}{4} =$ _____

Solve. Annex 0s as needed.

8.
$$5\overline{)21.1}$$

9.
$$8\overline{)15.6}$$

10.
$$2\overline{)29.63}$$

11.
$$6\overline{)65.1}$$

12.
$$7\overline{)29.806}$$

Write = or ≠ to compare.

13. $0.75 \bigcirc \frac{3}{4}$

14. $5.4 \bigcirc 5.04$

15. $0.75 \bigcirc \frac{7}{100}$

16. $\frac{6}{100} \bigcirc 0.06$

Solve. Label your answer.

17. Steve had 3.25 gallons of chocolate ice cream, 2.75 gallons of vanilla, and 3.0 gallons of strawberry. How much did he have altogether?

18. If Steve divided the chocolate ice cream into 5 equal parts, how much would each part be?

Solve. Check by multiplying.

1. $7 \overline{)33.6}$ Check

2. $5 \overline{)31.65}$ Check

3. $8 \overline{)27.96}$ Check

Write a.m. or p.m. for the activity.

4. School was over at 2:45 _____

5. Sunday school starts at 9:40 _____

6. Nathan and Laura watched the sunset at 7:30 _____

7. Brian had lunch at 12:00 _____

8. Lester set his alarm for 5:30 _____

Complete the table.

9.	**Standard form**	
10.	**Word form**	
11.	**Expanded form**	
12.	**Expanded form with multiplication**	$(4 \times 10{,}000{,}000) + (4 \times 1{,}000{,}000) + (8 \times 100{,}000) + (9 \times 10{,}000) + (3 \times 1{,}000) + (2 \times 10) + (1 \times 1)$

Multiply.

13. $\begin{array}{r} 3 \\ \times 4 \\ \hline \end{array}$

14. $\begin{array}{r} 8 \\ \times 7 \\ \hline \end{array}$

15. $\begin{array}{r} 6 \\ \times 5 \\ \hline \end{array}$

16. $\begin{array}{r} 9 \\ \times 7 \\ \hline \end{array}$

17. $\begin{array}{r} 4 \\ \times 6 \\ \hline \end{array}$

18. $\begin{array}{r} 6 \\ \times 8 \\ \hline \end{array}$

Divide.

19. $8 \overline{)64}$

20. $21 \div 3 = $ _____

21. $\frac{9}{3} = $ _____

22. $\frac{15}{5} = $ _____

23. $9 \overline{)72}$

24. $42 \div 7 = $ _____

25. $\frac{10}{2} = $ _____

26. $\frac{25}{5} = $ _____

Zero in the Quotient

Name _____

Solve.

1.
$$2\overline{)20.8}$$

2.
$$9\overline{)45.36}$$

3.
$$5\overline{)\$15.30}$$

4.
$$8\overline{)0.272}$$

Solve. Label your answer.

5. Mr. Jones bought $72.18 worth of candy over 9 days. How much on average did he spend each day?

_____

6. The Mendoza family spent $204.95 to rent a car for 5 days. How much did they spend for each day?

_____

7. Sandy ran 94.5 miles in 9 days. How many miles did she average each day?

_____

8. Lori read a total of 132.5 pages in 5 days. How many pages on average did she read each day?

_____

Solve. Check by multiplying.

9.
$$3\overline{)18.99}$$

Check

10.
$$4\overline{)\$17.20}$$

Check

11.
$$6\overline{)1.206}$$

Check

Round the decimal to the nearest one, tenth, and hundredth.

		one	tenth	hundredth
12.	34.568			
13.	30.718			

Write the decimal in fraction form.

14. $0.63 =$

15. $0.05 =$

16. $0.051 =$

17. $0.1 =$

Write the fraction in decimal form.

18. $\frac{4}{10} =$ _____

19. $\frac{33}{1,000} =$ _____

20. $\frac{52}{100} =$ _____

21. $\frac{4}{100} =$ _____

Solve.

1.
$$8\overline{)\$12.40}$$

2.
$$8\overline{)\$3.92}$$

3.
$$13\overline{)\$31.07}$$

Solve. Label your answer.

4. Sam traveled 248.4 miles. It took him 4 hours to get home. How many miles per hour was he traveling on average?

5. Sam's car used 9 gallons of gas for the trip. How many miles per gallon did he average?

Complete the table.

6.
pints	cups
2	
4	
6	
8	

7.
quarts	gallons
8	
16	
24	
32	

8.
pounds	ounces
4	
5	
7	
9	

9.
tons	pounds
1	
2	
4	
5	

Use the graph to answer the question.

The fifth graders at Grace Christian Elementary School made a graph to show how much money each grade raised for missions in the month of February.

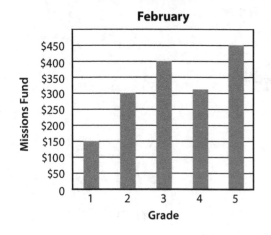

10. Which grade collected the most money?

_____

11. Which grade collected the least money?

_____

12. Which 2 grades collected about the same amount?

_____

13. How much more money did third grade collect than first grade?

_____

14. Which combined grades collected more: fourth and fifth grades or second and third grades?

_____

Powers of Ten

> When you multiply a decimal by 10 or a multiple of 10, the decimal point moves one place to the right for each 0 in the 10 factor. Annex a 0 when necessary.
> $10 \times 9.25 = 92.5$

Use mental math to solve.

1. $10 \times 5.3 =$ _____
2. $10 \times 11.63 =$ _____
3. $10 \times 0.082 =$ _____
4. $100 \times 7.1 =$ _____
5. $100 \times 21.02 =$ _____
6. $100 \times 1.011 =$ _____
7. $1,000 \times 4.2 =$ _____
8. $1,000 \times 9.14 =$ _____
9. $1,000 \times 8.095 =$ _____

Use mental math to solve.

10. $6.489 \times 10^1 =$ _____
11. $4.32 \times 10^1 =$ _____
12. $6.489 \times 10^2 =$ _____
13. $4.32 \times 10^2 =$ _____
14. $6.489 \times 10^3 =$ _____
15. $4.32 \times 10^3 =$ _____

Exponents
$10^1 = 10$
$10^2 = 100$
$10^3 = 1,000$

The exponent tells you how many places to move the decimal point.

> When you divide a decimal by 10 or a multiple of 10, the decimal point moves one place to the left for each 0 in the 10 divisor. Annex a 0 when necessary.
> $43.7 \div 100 = 0.437$

Use mental math to solve.

16. $68.2 \div 10 =$ _____
17. $344.88 \div 100 =$ _____
18. $227.44 \div 1,000 =$ _____
19. $42.01 \div 10 =$ _____
20. $334.1 \div 100 =$ _____
21. $809.2 \div 1,000 =$ _____

Use mental math to solve.

22. $63.2 \div 10^1 =$ _____
23. $63.2 \div 10^2 =$ _____
24. $63.2 \div 10^3 =$ _____
25. $0.4 \div 10^1 =$ _____
26. $0.4 \div 10^2 =$ _____
27. $0.4 \div 10^3 =$ _____

Solve. Label your answer.

28. MaryAnn babysits her brother for $3.50 per hour. How much money will she earn babysitting for 10 hours?

Solve.

1. Round 24.83 to the tenths place. _____

2. Which is larger: 27.1 or 27.01? _____

3. Who weighs more: Maggie at 98.4 pounds or Suzie at 90.8 pounds? _____

4. Round 1.809 to the Hundredths place. _____

5. Round 5.950 to the Ones place. _____

Write >, <, or = to compare.

6. $0.70 \bigcirc 0.700$ 7. $0.008 \bigcirc 0.8$ 8. $2.80 \bigcirc 2.08$

Solve. Rename if necessary.

9.   6 ft 8 in.   10.   3 ft 8 in.   11.   12 lb 14 oz   12.   4 tn 360 lb
   + 5 ft 3 in.   + 10 ft 6 in.   − 8 lb 10 oz   + 8 tn 720 lb

13.   3 ft 7 in.   14.   3 gal 2 qt   15.   3 hr 25 min   16.   4 hr 45 min
   − 2 ft 9 in.   + 9 gal 6 qt   + 2 hr 50 min   + 3 hr 30 min

Write the factors of the number. Label the number *prime* or *composite*.

17. 7: _____   18. 15: _____   19. 24: _____

_____   _____   _____

Solve.

20. $9 \cdot 9 =$ _____   21. $9 + 8 =$ _____   22. $7 + 9 =$ _____   23. $15 - 9 =$ _____

24. $5 \cdot 8 =$ _____   25. $18 - 10 =$ _____   26. $9\overline{)81}$ 27. $\frac{28}{7} =$ _____

Solving Problems

Solve. Show your work. Label your answer.

1. Mrs. Lippincott went shopping with wedding gift money. She bought two casserole dishes for $9.99 each, an electric mixer for $199.50, and 4 dish cloths for $2.50 each. She has $3.42 left. How much money had she received as a gift?

2. Gabrielle baked two and one-half times the number of cookies that Jenny baked. Jenny baked 48 cookies. How many cookies did Gabrielle bake?

3. How many dozen cookies did Gabrielle bake?

4. Brent was part of a marathon swim team raising money for a needy family. He swam 85.5 laps in 5 hours. How many laps did he average per hour?

5. Brent collected $4 for each lap that he swam. How much money did he collect altogether?

6. What is the average amount of money that Brent raised per hour? (He swam for 5 hours.)

Solve

The exponent tells you how many places to move the decimal point.

Multiplication: The decimal point moves to the right.

Division: The decimal point moves to the left.

Use mental math to solve. Annex 0s when needed.

1. $0.05 \times 10^2 =$ _____

2. $7.09 \times 10^1 =$ _____

3. $0.123 \times 10^3 =$ _____

4. $3.4 \times 10^1 =$ _____

5. $21.42 \times 10^3 =$ _____

6. $6.003 \times 10^2 =$ _____

7. $234.88 \div 10^2 =$ _____

8. $225 \div 10^1 =$ _____

9. $57.1 \div 10^3 =$ _____

10. $388 \div 10^3 =$ _____

11. $46.2 \div 10^2 =$ _____

12. $0.77 \div 10^1 =$ _____

Plot the mixed number on the number line. Round to the nearest whole number.

13. $2\frac{3}{5}$

14. $4\frac{5}{6}$

15. $1\frac{1}{6}$

16. $6\frac{4}{10}$

Add or subtract. Write the answer in lowest terms.

17. $\begin{array}{r} \frac{6}{10} \\ + \frac{8}{10} \\ \hline \end{array}$

18. $\begin{array}{r} 4\frac{10}{12} \\ + \frac{2}{12} \\ \hline \end{array}$

19. $\begin{array}{r} 3\frac{2}{4} \\ + 3\frac{2}{8} \\ \hline \end{array}$

20. $\begin{array}{r} \frac{2}{3} \\ + \frac{3}{7} \\ \hline \end{array}$

21. $\begin{array}{r} \frac{7}{7} \\ - \frac{6}{7} \\ \hline \end{array}$

22. $\begin{array}{r} 3 \\ - 1\frac{2}{7} \\ \hline \end{array}$

23. $\begin{array}{r} \frac{8}{10} \\ - \frac{3}{10} \\ \hline \end{array}$

24. $\begin{array}{r} \frac{3}{4} \\ - \frac{5}{8} \\ \hline \end{array}$

Solve.

25. $10 \times$ _____ $= 120$

26. $11 \times$ _____ $= 132$

27. _____ $\times 11 = 121$

28. _____ $\times 9 = 108$

29. $12 \times$ _____ $= 96$

30. _____ $\times 12 = 84$

Decimal Review

Name _____

Mark the answer.
Mark *NH* if the answer is "Not Here."

1. $\frac{2}{100} = \frac{?}{}$

   ○ 200   ○ 0.02
   ○ 0.200   ○ NH

2. $\frac{179}{1,000} = \frac{?}{}$

   ○ 0.179   ○ 0.0179
   ○ 1.79   ○ NH

3. $0.80 = \frac{?}{}$

   ○ $\frac{8}{100}$ ○ $\frac{80}{1,000}$
   ○ $8\frac{1}{10}$ ○ NH

4. $0.543 = \frac{?}{}$

   ○ $\frac{543}{100}$ ○ $5\frac{43}{100}$
   ○ $\frac{543}{1,000}$ ○ NH

5. five and twenty-two hundredths
   ○ 5.22   ○ 5.022
   ○ 0.522   ○ NH

6. one hundred forty-seven thousandths
   ○ 147,000   ○ 1.047
   ○ 100.47   ○ NH

7. the value of 5 in 5.371
   ○ 50   ○ 5
   ○ 0.5   ○ NH

8. place value of 5 in 46.58
   ○ tens   ○ tenths
   ○ ones   ○ NH

9. Round 2.761 to the nearest tenth.
   ○ 2.8   ○ 2.7
   ○ 2.76   ○ NH

10. Round 47.085 to the nearest hundredth.
    ○ 47.09   ○ 47.08
    ○ 47.1   ○ NH

11. Estimate the product of 4.176×3.
    ○ 15   ○ 12
    ○ 16   ○ NH

12. Estimate the product of 3.7×1.8.
    ○ 3   ○ 8
    ○ 4   ○ NH

13. $\begin{array}{r} 5.06 \\ \times\ \ 2 \\ \hline ? \end{array}$

    ○ 10.12
    ○ 1.012
    ○ 101.2
    ○ NH

14. $\begin{array}{r} 21.8 \\ \times\ 0.2 \\ \hline ? \end{array}$

    ○ 4.36
    ○ 0.436
    ○ 43.06
    ○ NH

15. $\begin{array}{r} 0.43 \\ \times 0.52 \\ \hline ? \end{array}$

    ○ 0.2236
    ○ 2.236
    ○ 22.36
    ○ NH

Solve. Label your answer.

16. If a gallon of gasoline costs $2.50 and George filled up his truck with 9.8 gallons, how much money did he spend on gas?

Mark the answer.
Mark *NH* if the answer is "Not Here."

17. $37.3 \div 4 = \underline{\ ?\ }$

    ○ 93.25   ○ 0.9325
    ○ 9.325   ○ NH

18. 60.84 meters of wire cut into 6 equal pieces

    ○ 11.14   ○ 10.14
    ○ 11.04   ○ NH

19. 1.5 pounds of butter for 3 cakes

    ○ 5 lb   ○ 2 lb
    ○ 0.05 lb   ○ NH

20. $\frac{1}{4} = \underline{\ ?\ }$

    ○ 0.25   ○ 4
    ○ 2.5   ○ NH

21. $\frac{1}{8} = \underline{\ ?\ }$

    ○ 0.125   ○ 12.5
    ○ 1.25   ○ NH

Solve. Check by multiplying.

22. $3\overline{)6.12}$

    Check

23. $5\overline{)357.5}$

    Check

Use mental math to solve.

24. $10 \times 4.2 = \underline{\hspace{2cm}}$

25. $100 \times 33.1 = \underline{\hspace{2cm}}$

26. $10^3 \times 5.143 = \underline{\hspace{2cm}}$

27. $83 \div 10^1 = \underline{\hspace{2cm}}$

28. $345.88 \div 1,000 = \underline{\hspace{2cm}}$

29. $116.3 \div 10^2 = \underline{\hspace{2cm}}$

Complete the table.

30.

Rule: ÷ 2	
Input	Output
14.8	
6.4	
22.6	
32.4	

31.

Rule: × 4	
Input	Output
1.4	
2.2	
3.6	
4.2	

Solve. Label your answer.

32. Dean ate 0.25 of the cookies that Faith baked. If Faith baked 48 cookies, how many did Dean eat?

33. Donna made 8 batches of peanut butter brownies. Each batch needed $\frac{2}{3}$ cup of peanut butter. How many cups of peanut butter did Donna need?

Name _____

Mark the answer.

1. Which equation shows the area of the rectangle?

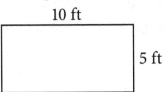

10 ft

5 ft

A. 10 ft × 5 ft = 50 ft²

B. 10 ft + 5 ft = 15 ft²

C. 10 ft ÷ 5 ft = 2 ft²

D. none of the above

2. Which equation shows the perimeter of the parallelogram?

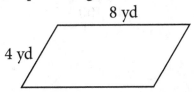

8 yd

4 yd

A. 4 yd + 8 yd = 12 yd

B. 4 yd × 8 yd = 32 yd

C. (2 × 8 yd) + (2 × 4 yd) = 24 yd

D. none of the above

3. How did the figure move?

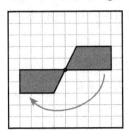

A. reflection

B. rotation

C. translation

4. What is the measure of the unknown angle of the triangle?

60°

? 60°

A. 50°   C. 70°

B. 60°   D. 80°

5. What is the measure of the unknown angle of the quadrilateral?

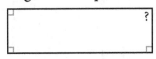

?

A. 70°   C. 90°

B. 80°   D. 100°

6. Which figure is congruent?

A.   C.

B.   D.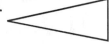

7. Which line segments are parallel?

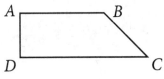

A   B

D   C

A. $\overline{AB}$ and $\overline{BC}$

B. $\overline{BC}$ and $\overline{AD}$

C. $\overline{AD}$ and $\overline{DC}$

D. $\overline{AB}$ and $\overline{DC}$

Mark the answer.

8. $\dfrac{8}{12} = \dfrac{?}{\underline{}}$

 A. $\dfrac{1}{2}$ C. $\dfrac{5}{6}$

 B. $\dfrac{2}{3}$ D. all of the above

9. $\dfrac{7}{8} + \dfrac{3}{4} = \dfrac{?}{\underline{}}$

 A. $1\dfrac{5}{8}$ C. $2\dfrac{1}{4}$

 B. 2   D. none of the above

10. $\dfrac{4}{9} + \dfrac{7}{9} = \dfrac{?}{\underline{}}$

 A. $\dfrac{1}{3}$ C. 1

 B. $\dfrac{8}{9}$ D. $1\dfrac{2}{9}$

11. $1\dfrac{2}{3} + 2\dfrac{1}{9} = \dfrac{?}{\underline{}}$

 A. $3\dfrac{1}{3}$ C. $3\dfrac{7}{9}$

 B. $3\dfrac{4}{9}$ D. $3\dfrac{2}{3}$

12. $\dfrac{10}{2} + \dfrac{6}{8} = \dfrac{?}{\underline{}}$

 A. $4\dfrac{7}{8}$ C. $6\dfrac{1}{4}$

 B. $5\dfrac{3}{4}$ D. $6\dfrac{1}{2}$

13. five-ninths more than $6\dfrac{4}{9} = \dfrac{?}{\underline{}}$

 A. 6   C. 7

 B. $6\dfrac{8}{9}$ D. $11\dfrac{4}{9}$

14. $(20 \times 200) + (20 \times 50) + (20 \times 7) = \dfrac{?}{\underline{}}$

 A. $4{,}000 + 100 + 140$

 B. $4{,}000 + 1{,}000 + 140$

 C. $400 + 100 + 140$

 D. $400 + 100 + 14$

15. $6 \times \dfrac{1}{3} = \dfrac{?}{\underline{}}$

 A. 1   C. 2

 B. $1\dfrac{1}{3}$ D. $2\dfrac{1}{3}$

16. Which is *not* a name for 38?

 A. 2×19 C. $(10 \times 2) + (9 \times 2)$

 B. $76 \div 2$ D. $100 - 78$

17.

n		
15	15	15

 A. $n = 30$ C. $n = 60$

 B. $n = 45$ D. $n = 90$

18. At the yard sale, Bryce bought a bicycle for $15, a helmet for $7, and 2 puzzles for $3.75. How much money does he have left from $40?

 A. $10.75   C. $14

 B. $12.25   D. $14.25

3-Dimensional Figures

Name _____

Use the word bank to complete the statement.

| height   length   width |

1. 2-dimensional figures have _____ and _____.

2. 3-dimensional figures have _____, _____, and _____.

Write *2-D* or *3-D* to name the shape.

3.
4.
5.
6.
7.
8.

_____   _____   _____   _____   _____   _____

Write the name of the figure that is described. For problems 10–13, draw a line from the name of the figure to its net.

| cone   cylinder   sphere   square prism   square pyramid |

9. no faces, no edges, no vertices   _____

10. one circular face, one curved surface, one vertex   _____

11. four triangular faces, one square face   _____

12. two circular faces, one curved surface   _____

13. six faces, eight vertices   _____

Write the correct word.

14. A(n) _____ is where three edges meet and is also the point of a cone.

15. A(n) _____ is a flat surface of a solid figure.

16. A(n) _____ is where two faces meet.

17. The flat surface of a cone is also called a(n) _____.

18. The flat pattern of a 3-dimensional figure is called a(n) _____.

| base |
| edge |
| face |
| net |
| vertex |

Use the line graph to answer the question.

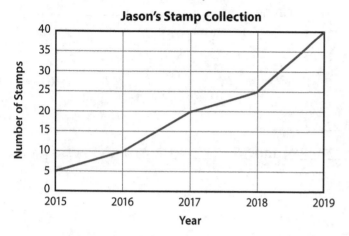

Jason's Stamp Collection

1. How many stamps did Jason have when he started his collection in 2015?

   _____

2. How many stamps did he have in 2019?

   _____

3. In what year did Jason have 25 stamps in his collection?

   _____

4. How many more stamps did Jason have in 2019 than 2018? _____

5. As time goes by, what is happening to the number of stamps in Jason's collection?

   _____

6. If this trend continues, how many stamps do you think Jason might have in 2020?

   _____

   Explain your answer.

   _____

Match the name of the property to the equation.

_____ 7. $2{,}142 \times 0 = 0$ A Associative

_____ 8. $15 \times 24 = (10 \times 24) + (5 \times 24)$ B Identity

_____ 9. $(30 \times 25) \times 4 = 30 \times (25 \times 4)$ C Commutative

_____ 10. $32 \times 25 = 25 \times 32$ D Distributive

_____ 11. $1 \times 607 = 607$ E Zero

Solve. Write the answer in lowest terms.

12. $\frac{2}{3} \times \frac{1}{3} =$ 13. $\frac{2}{5} \times \frac{3}{4} =$ 14. $\frac{4}{7} \times \frac{2}{3} =$ 15. $\frac{2}{5} \times \frac{2}{3} =$

16. $\frac{4}{5} \times \frac{5}{8} =$ 17. $\frac{3}{4} \times \frac{4}{7} =$ 18. $\frac{1}{4} \times \frac{1}{2} =$ 19. $\frac{1}{2} \times \frac{5}{6} =$

Write the product of the multiplication fact. Write a related division fact.

20. $3 \times 7 =$ _____   21. $6 \times 8 =$ _____   22. $4 \times 7 =$ _____

   _21_ ÷ _7_ = _____   _____ ÷ _____ = _____   _____ ÷ _____ = _____

23. $5 \times 9 =$ _____   24. $6 \times 6 =$ _____   25. $8 \times 7 =$ _____

   _____ ÷ _____ = _____   _____ ÷ _____ = _____   _____ ÷ _____ = _____

Prisms & Pyramids

Name _____

Write *rectangular*, *square*, or *triangular* to name the prism. Write the correct number of faces, vertices, and edges.

	Figure	Prism	Faces	Vertices	Edges
1.	bases				
2.	bases				
3.	bases				

Answer the question.

4. Which two prisms have the same number of faces, vertices, and edges?

   _____, _____

5. How is a prism named?

   _____

Write *rectangular*, *square*, or *triangular* to name the pyramid and its net.

6.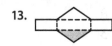

   _____ pyramid

7.

   _____ pyramid

8.

   _____ pyramid

Answer the question.

9. How many bases does a prism have? _____

10. How many bases does a pyramid have? _____

11. Which has at least three triangular faces: a prism or a pyramid? _____

12. Three triangular faces meet at a common point called a _____.

Write *prism* or *pyramid* to complete the name of the net.

13. triangular

    _____

14. 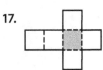 square

    _____

15. 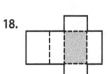 rectangular

    _____

16. triangular

    _____

17. square

    _____

18. rectangular

    _____

Use the pictograph to answer the question.

Cookie Sale	
Grade	Boxes of Cookies Sold
3rd	🍪 🍪 🍪 🍪
4th	🍪 🍪 🍪
5th	🍪 🍪
6th	🍪 🍪 🍪 🍪

🍪 = 10 boxes

1. Which grade sold the most cookies?

_____

2. Which grade sold twice as many cookies as 5th grade?

_____

3. What does 🍪 represent? _____

4. How many more boxes of cookies did 6th grade sell than 4th grade?

_____

Write a division equation to find the missing factor. Solve for *n*.

5. $7 \cdot n = 56$

$n =$ _____ ÷ _____

$n =$ _____

6. $6 \cdot n = 36$

$n =$ _____ ÷ _____

$n =$ _____

7. $8 \cdot n = 32$

$n =$ _____ ÷ _____

$n =$ _____

Write the quotient.

8. $\frac{63}{9} =$ _____

9. $\frac{36}{4} =$ _____

10. $\frac{42}{7} =$ _____

11. $\frac{49}{7} =$ _____

Divide.

12. $3\overline{)95}$

13. $2\overline{)2.34}$

14. $5\overline{)839}$

15. $5\overline{)6.15}$

Multiply.

16. $\begin{array}{r} 3.125 \\ \times \quad 5 \\ \hline \end{array}$

17. $\begin{array}{r} 2.32 \\ \times \ 0.6 \\ \hline \end{array}$

18. $\begin{array}{r} 4.1 \\ \times \ 3 \\ \hline \end{array}$

19. $\begin{array}{r} 0.78 \\ \times \ 0.3 \\ \hline \end{array}$

Surface Area

Name _____

Write the formula for the area of a 2-dimensional figure.

1. Area = _____ × _____

Write an equation to find the area.

2.

2 ft

5 ft

_____ ft × _____ ft = _____ ft²
length   width   area

3.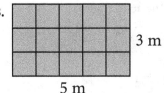

3 m

5 m

_____ m × _____ m = _____ m²

4.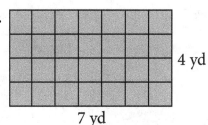

4 yd

7 yd

_____ yd × _____ yd = _____ yd²

Complete the equations to find the surface area of the rectangular prism.

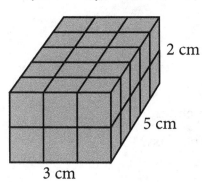

2 cm

5 cm

3 cm

5.

top = __5__ cm × __3__ cm = __15__ cm²
bottom = _____ cm × _____ cm = _____ cm²

front = _____ cm × _____ cm = _____ cm²
back = _____ cm × _____ cm = _____ cm²

right side = _____ cm × _____ cm = _____ cm²
left side = _____ cm × _____ cm = _____ cm²

6. (_____ cm² + _____ cm²) + (_____ cm² + _____ cm²) + (_____ cm² + _____ cm²) = _____ cm²
    top/bottom   front/back   sides   total surface area

Complete the equations to find the surface area of the cube.

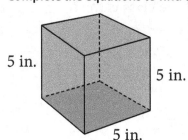

5 in.

5 in.

5 in.

> To find the total surface area of a cube, multiply the area of one face by 6.

7. What is the area of one face of the cube?

_____ in. × _____ in. = _____ in.²
length   width   area

8. What is the total surface area of the cube?

_____ × _____ = _____ in.²
# of faces area of one total surface
          face   area

9. Explain why the area of one face is multiplied by 6 to find the total surface area of the cube.

_____

Write *prism, pyramid,* or *both* for the description.

1. _____ at least two faces that are identical and parallel

2. _____ named by the shape of its base

3. _____ at least 3 triangular faces that meet at a common vertex

4. _____ has a polygon as its base

Write *rectangular, square,* or *triangular* to name the solid figure.

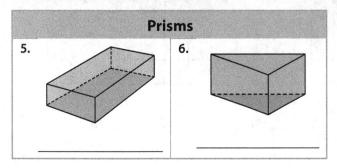

Prisms		Pyramids	
5.	6.	7.	8.

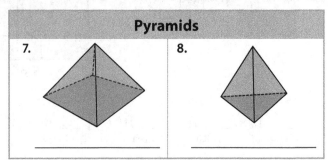

Use the circle graph to answer the question.

The fourth-grade class took a survey to find the favorite pizza toppings. The graph shows the results.

Favorite Pizza Toppings

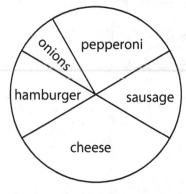

9. What fraction of the class likes pepperoni? _____

10. What is the favorite topping of the class?
_____

11. What two toppings did an equal number of students choose?
_____

12. Shade the parts of the graph representing the students that like cheese and and the students that like sausage. What fraction of the class is this? _____

Rename the fraction in lowest terms.

13. $\dfrac{4 \div \boxed{4}}{8 \div \boxed{4}} = \dfrac{}{}$

14. $\dfrac{5 \div \boxed{}}{10 \div } = \dfrac{}{}$

15. $\dfrac{3 \div \boxed{}}{9 \div } = \dfrac{}{}$

16. $\dfrac{9 \div \boxed{}}{12 \div } = \dfrac{}{}$

17. $\dfrac{10 \div \boxed{}}{20 \div } = \dfrac{}{}$

18. $\dfrac{7 \div \boxed{}}{21 \div } = \dfrac{}{}$

19. $\dfrac{4 \div \boxed{}}{24 \div } = \dfrac{}{}$

20. $\dfrac{15 \div \boxed{}}{36 \div } = \dfrac{}{}$

Divide. Write one related multiplication equation.

21. $81 \div 9 =$ _____

22. $32 \div 8 =$ _____

23. $28 \div 4 =$ _____

24. $72 \div 9 =$ _____

_____ _____ _____ _____

Volume

Write an equation to find the volume.

1.

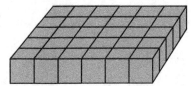

$\underline{\ \ 6\ \ }$ ft + $\underline{\ \ 5\ \ }$ ft + $\underline{\ \ 1\ \ }$ ft = $\underline{\ \ 30\ \ }$ ft³
   length   width   height   volume

2.

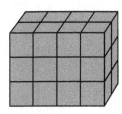

_____ cm + _____ cm + _____ cm = _____ cm³
  length   width   height   volume

3.

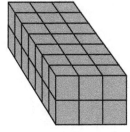

_____ m + _____ m + _____ m = _____ m³
  length   width   height   volume

4.

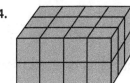

_____ yd + _____ yd + _____ yd = _____ yd³
  length   width   height   volume

Draw a line to match the term to its equation.

5. perimeter   = length × width × height

6. area   = length × width

7. volume   = side + side + side + side

What must you find to solve the following? Write _perimeter_, _area_, or _volume_.

8. amount of paint needed to paint a garage floor _____

9. amount of water needed to fill an aquarium _____

10. amount of fencing needed around a garden _____

Write an equation to find the perimeter and area.

6 ft

10 ft

11. $P =$ _____ ft + _____ ft + _____ ft + _____ ft = _____ ft

12. $A =$ _____ ft × _____ ft = _____ ft²

Complete the equations to find the surface area and volume.

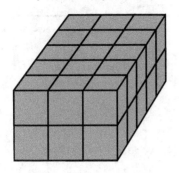

1.
top = _____ cm × _____ cm = _____ cm²
bottom = _____ cm × _____ cm = _____ cm²

front = _____ cm × _____ cm = _____ cm²
back = _____ cm × _____ cm = _____ cm²

right side = _____ cm × _____ cm = _____ cm²
left side = _____ cm × _____ cm = _____ cm²

2. (_____ cm² + _____ cm²) + (_____ cm² + _____ cm²) + (_____ cm² + _____ cm²) = _____ cm²
   top/bottom   front/back   sides   total surface area

3. $V =$ _____ cm × _____ cm × _____ cm = _____ cm³
      length   width   height   volume

4. How much wrapping paper is needed to wrap a box this size? _____

5. How much water would a fish tank this size hold? _____

Find the measure of the unknown angle.

Quadrilateral: a polygon with 4 sides $\angle A + \angle B + \angle C + \angle D = 360°$

6.
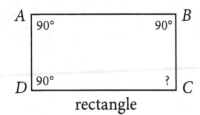
rectangle

_____ + _____ + n + _____ = 360°

_____ + n = 360°

n = _____

7.
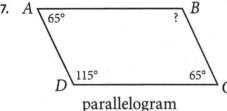
parallelogram

_____ + n + _____ + _____ = 360°

_____ + n = 360°

n = _____

Write an equation to find the area of the shaded triangle.

8.
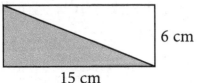
6 cm

15 cm

Area of ▭ = _____

Area of △ = _____

9.
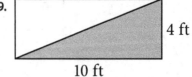
4 ft

10 ft

Area of ▭ = _____

Area of △ = _____

More Volume

Write an equation to find the perimeter and the area.

1. 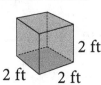 3 in.

5 in.

$P =$ _____ in. + _____ in. + _____ in. + _____ in. = _____ in.

$A =$ _____ in. × _____ in. = _____ in.^{2}

2. 7 cm

2.5 cm

$P =$ _____ cm + _____ cm + _____ cm + _____ cm = _____ cm

$A =$ _____ cm × _____ cm = _____ cm^2

Write an equation to find the volume.

3. 2 ft

2 ft   2 ft

_____

$V =$ _____ ft^3

4. 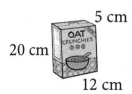 5 cm

20 cm

12 cm

_____

$V =$ _____ cm^3

5. 2 in.

10 in.

5 in.

_____

$V =$ _____

6. 4 in.

5 in.

6 in.

_____

$V =$ _____

Solve.

7. Phoebe the turtle needs clean water in her tank. The tank is 18 inches long, 10 inches wide, and 10 inches high. How much water will it take to fill the tank?

_____ in. × _____ in. × _____ in.

$V =$ _____ in.^{3}

8. Mrs. Bradstreet is going to plant some flowers to put on her patio. The planter is 4 feet long, 1 foot wide, and 2 feet high. How much potting soil will she need to fill the container?

_____ ft × _____ ft × _____ ft

$V =$ _____ ft^3

9. How many cubic feet of air are in a room that is 23 feet long, 18 feet wide, and 10 feet high?

_____

$V =$ _____

10. How much cereal will fill a container that is 5 inches wide, 12 inches long, and 10 inches high?

_____

$V =$ _____

Volume is the number of cubic units within a 3-dimensional figure.

$V = \underset{\text{length}}{\underline{4 \text{ units}}} \times \underset{\text{width}}{\underline{3 \text{ units}}} \times \underset{\text{height}}{\underline{2 \text{ units}}}$

$V = 24 \text{ units}^3$

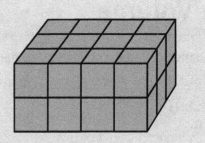

Write an equation to find the volume.

1.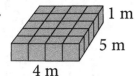
1 m
5 m
4 m

___4___ m × ___5___ m × ___1___ m

$V = $ _____ m³

2.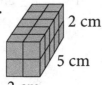
2 cm
5 cm
2 cm

_____ cm × _____ cm × _____ cm

$V = $ _____ cm³

3.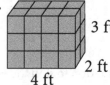
3 ft
2 ft
4 ft

_____

$V = $ _____

Write an equation to find the measure of the unknown angle.

The sum of the angle measures of any triangle is **180°**.

4.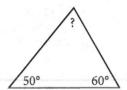
?
50°   60°

$180° - (\underline{\hspace{0.5cm}50°} + \underline{\hspace{0.5cm}60°}) = \underline{\hspace{0.5cm}?}$

_____ − _____ = _____

5.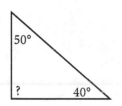
50°
?   40°

$180° - (\underline{\hspace{1cm}} + \underline{\hspace{1cm}}) = \underline{\hspace{0.5cm}?}$

_____ − _____ = _____

Write the fraction in decimal form.

6. $\frac{7}{10} = $ ___0.7___

7. $\frac{42}{100} = $ _____

8. $\frac{178}{1,000} = $ _____

9. $\frac{429}{1,000} = $ _____

10. $\frac{4}{100} = $ ___0.04___

11. $\frac{20}{100} = $ _____

12. $\frac{5}{100} = $ _____

13. $\frac{9}{1,000} = $ _____

Fill in the missing number.

14. $9 \times$ _____ $= 18$

15. $9 \times$ _____ $= 27$

16. $9 \times$ _____ $= 36$

17. $9 \times$ _____ $= 45$

18. _____ $\div 9 = 6$

19. _____ $\div 9 = 7$

20. _____ $\div 9 = 8$

21. _____ $\div 9 = 9$

22. $64 \div 8 = $ _____

23. $40 \div 8 = $ _____

24. $8 \times 7 = $ _____

25. $8 \times$ _____ $= 48$

26. _____ $\div 8 = 4$

27. _____ $\div 7 = 7$

28. $36 \div$ _____ $= 6$

29. $5 \times$ _____ $= 25$

More Surface Area & Volume

Name _____

To find **surface area**, find the sum of the areas of the faces.
To find **volume**, find the product of the length times the width times the height.

Complete the equations to find the surface area and volume.

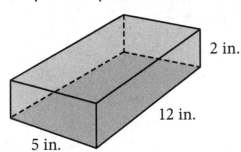

2 in.
12 in.
5 in.

1.

top = __12__ in. × __5__ in. = __60__ in.²
bottom = _____ in. × _____ in. = _____ in.²

front = _____ in. × _____ in. = _____ in.²
back = _____ in. × _____ in. = _____ in.²

right side = _____ in. × _____ in. = _____ in.²
left side = _____ in. × _____ in. = _____ in.²

2. (_____ in.² + _____ in.²) + (_____ in.² + _____ in.²) + (_____ in.² + _____ in.²) = _____ in.²
top/bottom   front/back   sides   total surface area

3. V = _____ in. × _____ in. × _____ in. = _____ in.³
     length   width   height   volume

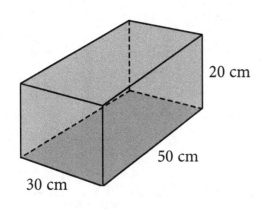

20 cm
50 cm
30 cm

4.

top = _____ cm × _____ cm = _____ cm²
bottom = _____ cm × _____ cm = _____ cm²

front = _____ cm × _____ cm = _____ cm²
back = _____ cm × _____ cm = _____ cm²

right side = _____ cm × _____ cm = _____ cm²
left side = _____ cm × _____ cm = _____ cm²

5. (_____ cm² + _____ cm²) + (_____ cm² + _____ cm²) + (_____ cm² + _____ cm²)
top/bottom   front/back   sides

= _____ cm²
total surface area

6. V = _____ cm × _____ cm × _____ cm = _____ cm³
     length   width   height   volume

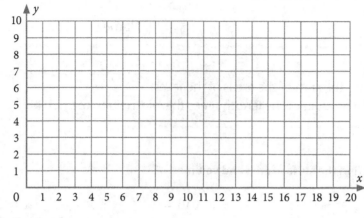

Plot and label the point on the graph.

1. *A* (3, 1)   2. *B* (8, 1)   3. *C* (3, 3)   4. *D* (8, 3)   5. *E* (4, 6)   6. *F* (8, 9)

7. *G* (9, 6)   8. *H* (3, 8)   9. *I* (12, 4)   10. *J* (19, 4)   11. *K* (15, 1)   12. *L* (15, 9)

Draw the line on the graph.

13. Draw $\overrightarrow{AB}$.   14. Draw $\overleftrightarrow{CD}$.   15. Draw $\overleftrightarrow{EF}$.   16. Draw $\overleftrightarrow{GH}$.   17. Draw $\overrightarrow{IJ}$.   18. Draw $\overleftrightarrow{KL}$.

Write *intersecting*, *parallel*, or *perpendicular* to classify the lines on the graph.

19. $\overleftrightarrow{AB}$ and $\overleftrightarrow{CD}$ are

_____.

20. $\overleftrightarrow{EF}$ and $\overleftrightarrow{GH}$ are

_____.

21. $\overleftrightarrow{IJ}$ and $\overleftrightarrow{KL}$ are

_____.

Use the word bank to complete the statement.

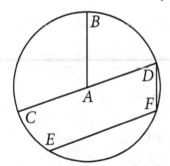

22. $\overline{AB}$ is a _____.

23. $\overline{CD}$ is a _____.

24. $\overline{EF}$ is a _____.

25. $\overline{DF}$ is a _____.

26. The name of this figure is _____ *A*.

circle

diameter

radius

chord

Write an equation to find the area.

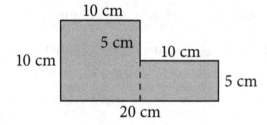

27. *A* = (_____ cm × _____ cm) + (_____ cm × _____ cm)

   shape 1   shape 2

A = _____ cm² + _____ cm²

A = _____ cm²

Complete the fact.

28. 1 minute = _____ seconds   29. 1 year = _____ days   30. 1 century = _____ years

31. 1 hour = _____ minutes   32. 1 year = _____ weeks   33. 1 millennium = _____ years

Geometry Review

Name _____

Mark the answer.
Mark *NH* if the answer is "Not Here."

1.
   - ○ cylinder
   - ○ cone
   - ○ sphere
   - ○ NH

2.
   - ○ cone
   - ○ triangular prism
   - ○ triangular pyramid
   - ○ NH

3.
   - ○ square prism
   - ○ square pyramid
   - ○ rectangular pyramid
   - ○ NH

4.
   - ○ cone
   - ○ prism
   - ○ cylinder
   - ○ NH

5.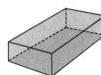
   - ○ rectangular pyramid
   - ○ cylinder
   - ○ cube
   - ○ NH

6.
   - ○ triangular prism
   - ○ square pyramid
   - ○ cone
   - ○ NH

7.
   - ○ triangular prism
   - ○ triangular pyramid
   - ○ cylinder
   - ○ NH

8.
   - ○ triangular prism
   - ○ triangular pyramid
   - ○ cone
   - ○ NH

9.
   - ○ cone
   - ○ cylinder
   - ○ sphere
   - ○ NH

10.
    - ○ square prism
    - ○ square pyramid
    - ○ cone
    - ○ NH

11.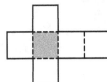
    - ○ sphere
    - ○ pyramid
    - ○ cone
    - ○ NH

12.
    - ○ sphere
    - ○ square pyramid
    - ○ square prism
    - ○ NH

13.
    - ○ triangular prism
    - ○ cylinder
    - ○ cone
    - ○ NH

14.
    - ○ cylinder
    - ○ rectangular prism
    - ○ rectangular pyramid
    - ○ NH

Find the perimeter of the figure. Mark *NH* if the answer is "Not Here."

15.
5 in.
5 in.

○ 20 in.
○ 25 in.
○ 25 in.²
○ NH

16.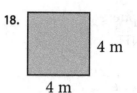
4 in.
8 in.

○ 32 in.²
○ 12 in.
○ 24 in.
○ NH

17.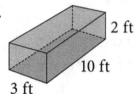
6 ft
3 ft

○ 18 ft²
○ 18 ft
○ 9 ft²
○ NH

18.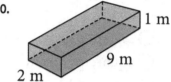
4 m
4 m

○ 16 m
○ 8 m
○ 12 m²
○ NH

Write an equation to find the volume.

19.
2 ft
10 ft
3 ft

V = _____

V = _____

20.
1 m
9 m
2 m

V = _____

V = _____

Find the surface area of the prism.

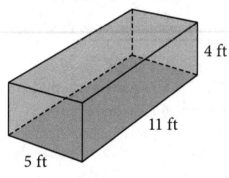
4 ft
11 ft
5 ft

21.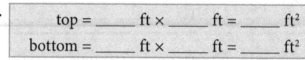

top = _____ ft × _____ ft = _____ ft²
bottom = _____ ft × _____ ft = _____ ft²

front = _____ ft × _____ ft = _____ ft²
back = _____ ft × _____ ft = _____ ft²

right side = _____ ft × _____ ft = _____ ft²
left side = _____ ft × _____ ft = _____ ft²

22. (_____ ft² + _____ ft²) + (_____ ft² + _____ ft²) + (_____ ft² + _____ ft²) = _____ ft²

Find the perimeter and area of the figure.

23.
4 m
4 m

P = _____

A = _____

24.
8 in.
12 in.

P = _____

A = _____

Mark the answer.

1.
$$2{,}396 \times 3{,}487 = 3{,}487 \times \underline{}$$

A. 2,396   C. 3,487

B. 2,478   D. 3,488

2.
$$78 \times 53 = 53 \times (\underline{} + 8)$$

A. 30   C. 70

B. 50   D. 78

3.
$$(64 \times 25) \times 10 = (64 \times \underline{}) \times 25$$

A. 4   C. 10

B. 5   D. 14

Mark the value of the expression if $n = 5$.

4.
$$(40 \div n) \cdot 10 = \underline{}$$

A. 40   C. 120

B. 80   D. none of the above

5.
$$6 \cdot n + 320 = \underline{}$$

A. 290   C. 350

B. 315   D. none of the above

6.
$$(125 \div n) \cdot 10 = \underline{}$$

A. 250   C. 400

B. 325   D. none of the above

Mark the operation that makes the equation true.

7. $(6 \times 7) \bigcirc 3 = 39$

A. +   C. ×

B. −   D. ÷

8. $30 \div (10 \bigcirc 7) = 10$

A. +   C. ×

B. −   D. ÷

9. $(64 \bigcirc 8) \times 7 = 56$

A. +   C. ×

B. −   D. ÷

10. $(1{,}000 \bigcirc 10) + 4 = 104$

A. +   C. ×

B. −   D. ÷

11. $500 \bigcirc (10 \times 20) = 300$

A. +   C. ×

B. −   D. ÷

12. $(42 \div 7) \bigcirc 6 = 36$

A. +   C. ×

B. −   D. ÷

Mark the answer.

13. $45\overline{)900}$

    A. 10   C. 15 r15

    B. 11 r5   D. 20

14. $25\overline{)759}$

    A. 30   C. 31

    B. 30 r9   D. 32

15. $39\overline{)4,567}$

    A. 97 r21   C. 111 r50

    B. 110 r15   D. 117 r4

16. $\begin{array}{r} 127 \\ \times\ 63 \\ \hline \end{array}$

    A. 7,901   C. 8,010

    B. 8,001   D. 8,901

17. $\begin{array}{r} 208 \\ \times\ 43 \\ \hline \end{array}$

    A. 8,641   C. 8,834

    B. 8,647   D. 8,944

18. $\begin{array}{r} 378 \\ \times 100 \\ \hline \end{array}$

    A. 3,780   C. 378,000

    B. 37,800   D. 370,800

19. $\begin{array}{l} 3 \text{ hr } 37 \text{ min} \\ + 2 \text{ hr } 45 \text{ min} \\ \hline \end{array}$

    A. 5 hr 52 min   C. 6 hr 45 min

    B. 6 hr 22 min   D. 7 hr 15 min

20. $\begin{array}{l} 6 \text{ ft } \ \ 9 \text{ in.} \\ -1 \text{ ft } 11 \text{ in.} \\ \hline \end{array}$

    A. 4 ft 10 in.   C. 5 ft 20 in.

    B. 5 ft 2 in.   D. 6 ft 20 in.

21. If $\frac{10}{10} = 1$ whole, what does $\frac{30}{10}$ equal?

    A. 2 wholes   C. 4 wholes

    B. 3 wholes   D. 5 wholes

22. What is the standard form for sixteen million, fifty-seven thousand, three hundred ninety-nine?

    A. 16,573,990   C. 1,653,399

    B. 16,057,399   D. 10,657,390

23. Mr. Fisher set up 26 rows of 15 chairs for the spring concert. He set up an additional 30 chairs the night of the concert. If all but 12 seats were used for the concert guests, how many people attended the concert?

    A. 396   C. 408

    B. 400   D. 420

Metric Measurement: Linear

Name _____

Write the unit you would use to measure: *mm, cm, m,* or *km*.

1. length of a car _____
2. distance to work _____
3. length of a hand _____
4. thickness of a nickel _____
5. height of a flagpole _____
6. distance to the moon _____

Match.

_____ 7. one-hundredth of a meter

_____ 8. ten meters

_____ 9. one-thousandth of a meter

_____ 10. one thousand meters

_____ 11. one hundred meters

_____ 12. one-tenth of a meter

A millimeter (mm)

B centimeter (cm)

C decimeter (dm)

D dekameter (dkm)

E hectometer (hm)

F kilometer (km)

Use a ruler to find the length in centimeters or millimeters.

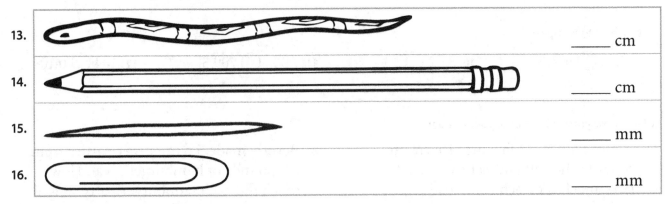

13. _____ cm

14. _____ cm

15. _____ mm

16. _____ mm

Solve.

> The boundaries of a farm measure 872 meters on the north side, 1 kilometer 50 meters on the south side, 703 meters on the east side, and 723 meters on the west side.

17. Which side is the longest? _____
18. Which side is the shortest? _____
19. How many meters long is the south side?

_____

20. How many meters is the perimeter of the farm? _____

> Three boys had a jumping contest. Aaron jumped 2 meters. Noah jumped 1 meter 75 centimeters, and William jumped 125 centimeters.

21. Who jumped the longest distance?

_____

22. How many centimeters did Noah jump?

_____

23. How many centimeters did Aaron jump?

_____

24. How many meters did the 3 boys jump altogether? _____

Match.

_____ 1.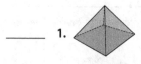

A cylinder

B square pyramid

C rectangular prism

_____ 2.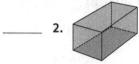

_____ 3.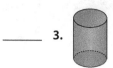

_____ 4.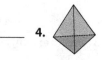

A cone

B triangular pyramid

C triangular prism

_____ 5.

_____ 6.

Write the decimals from _least_ to _greatest_.

7.

| 2.87 | 28.7 | 0.287 | 0.278 |

_____ _____ _____ _____

8.

| 5.42 | 5.4 | 5.24 | 5.37 |

_____ _____ _____ _____

Write > or < to compare.

9. 62.13 ◯ 6.213

10. 3.02 ◯ 3.1

11. 7.085 ◯ 709.5

12. 104 ◯ 0.104

Solve. Show your work. Label your answer.

13. Alyssa paid $7.83 for three ice cream cones for herself and her two friends. How much did each cone cost?

14. Beverly made 5 hamburger patties from 2.7 pounds of hamburger meat. How much did each hamburger patty weigh?

Write the equivalent measure.

15. _____ cm = 1 m

16. 1 dkm = _____ m

17. _____ mm = 1 m

18. 1 hm = _____ m

19. 1,000 m = _____ km

20. _____ dm = 1 m

More Linear Measurement
Name _____

> Multiply to rename larger units as smaller units.
> Divide to rename smaller units as larger units.

Rename the given measurement.

1. Jason rode his bike 4 kilometers to school. How many meters did he ride? _____

2. Susie walked the track for 2,000 meters. How many kilometers did she walk? _____

3. The bedroom is 5 meters wide. How many centimeters wide is the bedroom? _____

4. The table is 200 centimeters long. How many meters long is the table? _____

Write >, <, or = to compare. Rename the smaller unit if necessary.

5. 6 km $\bigcirc$ 5,600 m

6. 3,960 mm $\bigcirc$ 5 m

7. 4 cm $\bigcirc$ 44 mm

8. 9,000 m $\bigcirc$ 9 km

9. 80 cm $\bigcirc$ 8 m

10. 3 m $\bigcirc$ 300 mm

Multiply to rename centimeters as millimeters.

11. 7.9 cm = _?_ mm

> 1 cm = 10 mm

7.9 × __10__ = _____ mm

12. 23 cm = _?_ mm

23 × _____ = _____ mm

13. 83.4 cm = _?_ mm

83.4 × _____ = _____ mm

Divide to rename millimeters as centimeters.

14. 86 mm = _?_ cm

> 10 mm = 1 cm

86 ÷ __10__ = _____ cm

15. 743 mm = _?_ cm

743 ÷ _____ = _____ cm

16. 100 mm = _?_ cm

100 ÷ _____ = _____ cm

Write the unit you would use to measure: *m*, *cm*, or *mm*.

17. length of a pool _____

18. diameter of a penny _____

19. width of your foot _____

20. length of your school's hallway _____

Complete the equations to find the surface area of the rectangular prism.

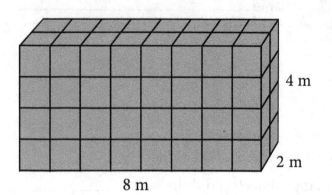

8 m

4 m

2 m

1.

top = _____ m × _____ m = _____ m²
bottom = _____ m × _____ m = _____ m²

right side = _____ m × _____ m = _____ m²
left side = _____ m × _____ m = _____ m²

front = _____ m × _____ m = _____ m²
back = _____ m × _____ m = _____ m²

2. (_____ m² + _____ m²) + (_____ m² + _____ m²) + (_____ m² + _____ m²) = _____ m²
    top/bottom   sides   front/back   total surface
                                                                                    area

Multiply. Count the decimals to find the decimal placement.

3. 1.4
   ×1.8

4. 2.8
   ×1.9

5. 0.36
   × 1.4

6. 0.4
   ×0.7

7. 32.9
   × 1.2

8. 0.26
   × 0.7

9. Find the product of 0.6 and 0.4. _____

10. Find the product of 1.83 and 2.1. _____

Use mental math to solve.

11. $10 \times 4.3 =$ _____

12. $100 \times 3.98 =$ _____

13. $1,000 \times 8.02 =$ _____

14. $49.2 \div 10 =$ _____

15. $1.03 \div 10 =$ _____

16. $5.63 \div 10 =$ _____

Complete the table.

17.

cm	m
100	
	3
800	

18.

mm	m
1,000	
4,000	
	6

19.

cm	mm
1	
5	
	90

Metric Measurement: Capacity & Mass

Name _____

Write the unit you would use to measure: *mL* or *L*.

1. _____ punch for a party
2. _____ water in a pool
3. _____ eye drops
4. _____ vanilla flavoring for a cake
5. _____ $\frac{1}{2}$ cup of milk
6. _____ fuel for a car

Rename the units.

7. 5 L = _____ mL
8. 500 mL = _____ L
9. 6,611 mL = _____ L
10. 4,000 mg = _____ g
11. 26.382 g = _____ mg
12. 126 mg = _____ g

Use the information from the chart to find the answer.

Type of Fruit Juice	Volume
orange	2,000 mL
pineapple	1,718 mL
apple	3 L
grape	1.6 L

13. What is the volume of the apple juice in milliliters?

_____

14. What is the volume of the orange juice in liters? _____

15. What is the volume of the grape juice in milliliters?

_____

16. What is the volume of the pineapple juice in liters?

_____

Write >, <, or = to compare.

17. 8,000 g ◯ 80 kg
18. 6,899 mg ◯ 6.899 g
19. 1,085 mg ◯ 0.1085 g

Solve.

20. 6 L – 1,000 mL = _____ L
21. 5 L + 2,640 mL = _____ mL
22. 4 kg + 300 g = _____ g
23. 7 kg – 2,000 g = _____ kg

Write the unit you would use to measure: *mg*, *g*, or *kg*.

24. _____ a baby
25. _____ an apple
26. _____ a grain of rice
27. _____ a spoon
28. _____ a whole watermelon
29. _____ a raindrop

Write the unit you would use to measure: *kilometer, meter, centimeter,* or *millimeter.*

1.
| thickness of pencil lead |

_____

2.
| distance from Washington to New York |

_____

3.
| length of a bicycle |

_____

4.
| length of your notebook |

_____

Mark the best estimate.

5.
| distance to church |

○   ○   ○
10 km   10 m   100 cm

6.
| length of a pencil |

○   ○   ○
20 mm   200 m   20 cm

Use a ruler to find the length in centimeters or millimeters.

7. ⊢————————————⊣ _____ cm

8. ⊢—————⊣ _____ mm

Write an equation to find the volume. Solve.

> Volume = length × width × height

9.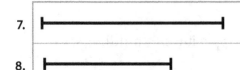

____ cm × ____ cm × ____ cm

$V =$ _____ cm^3

10.

____ m × ____ m × ____ m

$V =$ _____ m^3

11.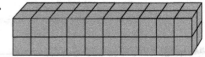

____ cm × ____ cm × ____ cm

$V =$ _____ cm^3

Use the line graph to answer the questions.

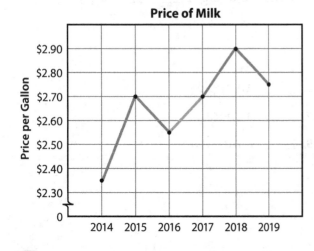

Price of Milk

12. What year did milk cost the most?

_____

13. Was there an increase or decrease in the cost of milk between 2016 and 2018? _____

How much? _____ _____

14. What was the lowest price for milk during the years 2014 to 2019? _____

15. What was the difference in price from 2014 to 2019? _____

Celsius Temperature

Name _____

Write the Celsius temperature shown. Write the new temperature.
Shade the second thermometer to show the new temperature.

1.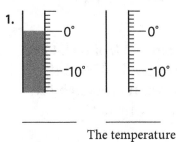

   _____   _____

   The temperature
   dropped 12°.

2.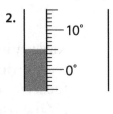

   _____   _____

   8° warmer

3.

   _____   _____

   7° colder

Mark the best estimate.

4.	bottle of soda	○ 20 L	○ 2 mL	○ 2 L
5.	a slice of toast	○ 30 kg	○ 30 g	○ 30 mg
6.	distance from Miami to Atlanta	○ 1,000 m	○ 1,000 km	○ 1,000 mm
7.	cup of hot tea	○ 250 mL	○ 250 L	○ 25 L
8.	the height of the ceiling	○ 30 cm	○ 30 m	○ 3 m
9.	a tiger	○ 250 g	○ 250 kg	○ 25 g
10.	water in a wading pool	○ 1,000 L	○ 1,000 mL	○ 10 L
11.	one egg	○ 50 kg	○ 50 g	○ 50 mg
12.	the length of a mosquito	○ 1 m	○ 5 mm	○ 50 cm

Mark the more reasonable temperature.

13.
summer picnic

   ○ 5°C
   ○ 25°C

14.
building a snowman

   ○ 2°C
   ○ 35°C

15.
freezing water

   ○ 0°C
   ○ 32°C

Solve. Write the equivalent.

16. 1,461 cm
   +   539 cm
   _____

   _____ cm = _____ m

17.   6 km
   − 2,000 m
   _____

   _____ m = _____ km

18. 4,000 g
   +   5 kg
   _____

   _____ g = _____ kg

Write >, <, or = to compare.

1. 4,862 mL ◯ 4 L

2. 7 L ◯ 7,000 mL

3. 350 mL ◯ 35 L

4. 6,000 g ◯ 60 kg

5. 17,302 g ◯ 17.302 kg

6. 1,000 mg ◯ 2 g

Solve.

7. Tim needs 5 kg of beef for the barbeque on Saturday. He bought one package with 3 kg and another with 1 kg 850 g. Does he have enough for the barbeque? Why or why not?

8. Karl's minivan holds 77 L of fuel. On a recent trip, the van used 80,000 mL of fuel. Was one tank of fuel enough fuel for the trip? Why or why not?

Write the numbers from *least* to *greatest*.

9.
| 38.2 | 38.02 | 3.802 |

_____ _____ _____

10.
| 98.32 | 9.823 | 983.2 |

_____ _____ _____

11.
| 6.153 | 6.943 | 6.099 |

_____ _____ _____

> The fifth-grade classes competed to recycle the most aluminum cans.

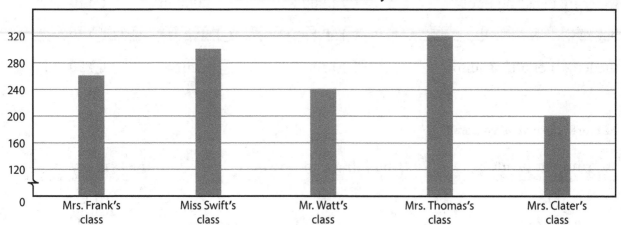

Pounds of Aluminum Recycled

Use the graph to answer the question.

12. Do the numbers on the scale count by 20, 30, 40, or 50? _____

13. Which class recycled 260 pounds of aluminum? _____

14. Which class recycled 300 pounds of aluminum? _____

15. Which class recycled 120 pounds more than Mrs. Clater's class? _____

16. Which class recycled the most aluminum? _____

Adding & Subtracting Metric Units

Name _____

Rename the metric units.

1. 72 m 10 cm = _____ cm

2. 8 L 75 mL = _____ mL

3. 5 kg 18 g = _____ g

4. 1,023 cm = _____ m

Solve. Rename if needed.

5. 7 kg 238 g
   + 4 kg 782 g

6. 5 cm 6 mm
   − 3 cm 14 mm

7.   7.05 L
   + 6.32 L

8. 37 m 19 cm
   − 14 m 23 cm

9. 9 L 374 mL
   + 3 L 626 mL

10. 10 m 81 cm
   + 9 m 35 cm

11.   4,427 L
   − 3,263 L

12. 17 kg 205 g
   − 8 kg 98 g

Use the table to find the answer.

Tallest Buildings in the United States	
30 Hudson Yards	387 m
432 Park Ave.	426 m
Aon Center	346 m
Bank of America Tower	366 m
Empire State Building	381 m
One World Trade Center	541 m
Trump International Hotel & Tower Chicago	423 m
Willis Tower	443 m

13. What is the height of the Empire State Building?

   _____

14. What is the height of 30 Hudson Yards in centimeters?

   _____

15. List the heights of the three tallest buildings.

   _____

16. Name the tallest and shortest buildings on the list. Give the difference in their heights.

   tallest: _____

   shortest: _____

   difference: _____

17. Plans are in progress to build a tower 1 kilometer tall. How many meters high will this building be when completed?

   _____

18. How much taller will the 1-kilometer tower be than the Willis Tower?

   _____

Write the Celsius temperature.

1. 20° 10°

_____

2. 0° ‾10°

_____

3. 0° ‾10°

_____

4.  20° 10°

_____

5. 0° ‾10°

_____

Write the Celsius temperature shown. Write the new temperature.
Shade the second thermometer to show the new temperature.

6. 0° ‾10°   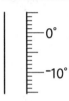 0° ‾10°

_____   _____

7° colder

7.  25° 15°    25° 15°

_____   _____

15° warmer

8. 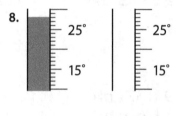 25° 15°   25° 15°

_____   _____

The temperature
dropped 11°.

Solve. Rename if needed.

9. 38 m 95 cm
   + 52 m 82 cm

10. 5 cm 6 mm
    − 2 cm 14 mm

11. 6 L 360 mL
    − 2 L 600 mL

12. 6 kg 238 g
    + 4 kg 782 g

Write the equivalent measure.

13. 1 cm = _____ mm

14. 1 m = _____ mm

15. 1 L = _____ mL

16. 1,000 g = _____ kg

17. 1 kg = _____ g

18. 1 g = _____ mg

19. 1 km = _____ m

20. 1 m = _____ cm

Rename the units.

21. 4 km = _____ m

22. 5 kg = _____ g

23. 8,000 mL = _____ L

24. 6 L = _____ mL

25. 3 cm = _____ mm

26. 2,000 g = _____ kg

Metric Measurement Review

Name _____

Mark the best estimate.

1. thickness of a nickel
   - ○ 20 mm
   - ○ 2 m
   - ○ 2 mm
   - ○ 2 cm

2. glass of water
   - ○ 350 L
   - ○ 35 L
   - ○ 35 mL
   - ○ 350 mL

3. a chocolate caramel
   - ○ 8 g
   - ○ 8 mg
   - ○ 8 kg
   - ○ 80 g

4. length of a bicycle
   - ○ 2 km
   - ○ 20 cm
   - ○ 2 m
   - ○ 200 mm

5. cup of coffee
   - ○ 250 L
   - ○ 250 mL
   - ○ 25 mL
   - ○ 25 L

6. distance to church
   - ○ 22 m
   - ○ 200 cm
   - ○ 2,000 mm
   - ○ 22 km

Mark the larger measurement.

7. ○ 14 mm or ○ 1 cm

8. ○ 10 km or ○ 1,000 cm

9. ○ 145 g or ○ 145 kg

10. ○ 3,000 mL or ○ 2 L

11. ○ 990 mg or ○ 1 kg

Mark the answer.

12. 742 cm = ___?___ m
    - ○ 7.42
    - ○ 0.0742
    - ○ 700.42
    - ○ 74.2

13. 2,807 g = ___?___ kg
    - ○ 28.07
    - ○ 0.2807
    - ○ 280.7
    - ○ 2.807

14. 62 cm = ___?___ mm
    - ○ 6.20
    - ○ 0.62
    - ○ 620
    - ○ 60.2

15. 5 L = ___?___ mL
    - ○ 0.005
    - ○ 5,000
    - ○ 50
    - ○ 500

Write the Celsius temperature shown.
Write the new temperature. Shade the second thermometer to show the new temperature.

16.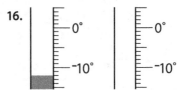

    _____   _____

    The temperature rose 12°.

17.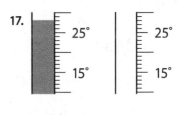

    _____   _____

    12° colder

Solve. Rename if necessary.

18. 16 kg 426 g
    − 7 kg 207 g

19. 10 L 342 mL
    + 2 L 419 mL

20.   6 m 98 cm
    + 2 m 14 cm

21.   7 m 83 cm
    − 5 m 98 cm

22. 19 kg 32 g = _____ g

23. 43 m 12 cm = _____ cm

24. 7,000 mL = _____ L

25. 4,000 mm = _____ m

Solve. Label your answer.

26. Eric collected 1,745 grams more candy
    than Timothy collected. Timothy
    collected 846 grams of candy. How much
    candy did Eric collect?

27. How many kilograms of candy did
    Timothy and Eric collect combined?

28. Meredith bought 2 watermelons. The
    larger melon weighed 3 kilograms. The
    smaller melon weighed 750 grams less
    than the larger one. What was the weight
    of the smaller one?

29. Jamie has a pet lizard and a pet snake.
    The lizard is 10 centimeters long. The
    snake is 1 meter long. How much longer is
    the snake than the lizard (in centimeters)?

30. Devin is 1.5 meters tall. What is Devin's
    height in centimeters?

Mark the answer.

1. Round 1,563,789 to the nearest one million.
   A. 1,564,000   C. 1,600,000
   B. 1,560,000   D. 2,000,000

2. Round 17.657 to the nearest tenth.
   A. 17.7   C. 18.6
   B. 17.66   D. 18.66

3. What is two thousand more than 529,631?
   A. 527,631   C. 531,631
   B. 529,831   D. 539,631

4. $6 \times 9 = (3 \times 15) + \underline{?}$
   A. 9   C. 20
   B. 12   D. none of the above

5. $(318 + 56) \times 0 < \underline{?} \times 1$
   A. 21   C. 420
   B. 180   D. all of the above

6.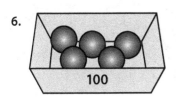
   A. 15   C. 25
   B. 20   D. 40

7. $3 \times 1.18 = \underline{?}$
   A. 3.08   C. 3.54
   B. 3.18   D. 13.60

8. $\frac{504}{24} = \underline{?}$
   A. 18   C. 25
   B. 21   D. 30

Sydney volunteers 6 hours each Saturday at a horse farm.

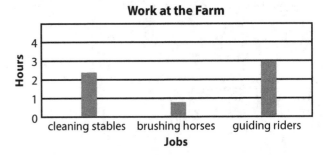

Work at the Farm

Use the bar graph to find the answer.

9. On which job does Sydney spend $\frac{1}{2}$ of her time?
   A. cleaning stables
   B. brushing horses
   C. guiding riders

10. About how much time does Sydney spend cleaning the stables?
    A. $2\frac{1}{4}$ hours
    B. $2\frac{3}{4}$ hours
    C. 3 hours

Mark the answer.

11.
$$5\overline{)15.3}$$ with $?$ above

A. 3.01   C. 3.1

B. 3.06   D. 3.6

12.
$367 \times 56 = \underline{\ ?\ }$

A. 2,202   C. 19,952

B. 10,552   D. 20,552

13. Estimate the difference to the nearest thousand: $21{,}293 - 8{,}946$.

A. 7,000   C. 9,000

B. 8,000   D. 12,000

14. Estimate the sum: $12\frac{5}{6} + 11\frac{1}{3}$.

A. 23   C. 24

B. $23\frac{1}{6}$ D. $24\frac{3}{4}$

15.
$\frac{6}{9} + \frac{3}{6} = \underline{\ ?\ }$

A. $\frac{9}{15}$ C. $1\frac{1}{6}$

B. $\frac{17}{18}$ D. $1\frac{2}{3}$

16.
$\frac{1}{5} + \frac{1}{3} = \underline{\ ?\ }$

A. $\frac{2}{8}$ C. $\frac{8}{15}$

B. $\frac{2}{5}$ D. $1\frac{1}{5}$

17.
3 sets of $\frac{2}{3}$ is $\underline{\ ?\ }$.

A. $1\frac{2}{3}$ C. $2\frac{1}{3}$

B. 2   D. 3

18.
$300 + 90 + 6 = \underline{\ ?\ }$

A. 132×3 C. $(10 + 30) \times 3$

B. 200×2 D. $400 - 10$

19.
$25 = \underline{\ ?\ }$

A. 17.5×2 C. $2 \times (5 + 5)$

B. 5^2 D. $2 \cdot 2 \cdot 2 \cdot 2 \cdot 2$

20.
$140 \times 63 = \underline{\ ?\ }$

A. $(20 \times 7) + 63$ C. $(20 + 7) \times (7 + 9)$

B. $100 \times (60 + 3)$ D. $(20 \times 7) \times (7 \times 9)$

21. How many faces are on the prism?

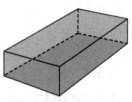

A. 4   C. 8

B. 6   D. 10

22. What equation shows the volume of the prism?

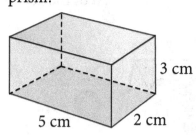

3 cm

5 cm   2 cm

A. 5 cm × 2 cm × 3 cm = 30 cm³

B. 5 cm + 2 cm + 3 cm = 10 cm³

C. (5 cm + 2 cm) × 3 cm = 25 cm³

D. 5 cm × (2 cm + 3 cm) = 16 cm³

Ratios

Write the ratio in each form.

> Mark's family owns 8 pets—3 hamsters and 5 guinea pigs.

	Ratio	Word Form	Ratio Form	Fraction Form
1.	hamsters to guinea pigs	3 to 5	3 : 5	$\frac{3}{5}$
2.	hamsters to pets	to 8	: 8	$\frac{}{8}$
3.	pets to hamsters	to	:	—
4.	guinea pigs to hamsters	5 to 3	:	—
5.	guinea pigs to pets	to	:	—
6.	pets to guinea pigs	to	:	—

Use the group of shapes to find the ratio. Write the ratio in three forms.

7. circles to triangles

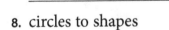

8. circles to shapes

_____

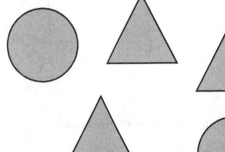

9. shapes to circles

_____

Use the shapes to write the ratio in words.

10.   4 to 3   4 : 3 $\frac{4}{3}$ triangles to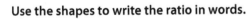

11.   4 to 7   4 : 7 $\frac{4}{7}$ _____

12.   7 to 4   7 : 4 $\frac{7}{4}$ _____

Complete the chart.

1.

Metric Units of Length						
kilometer (km)	hectometer (hm)	dekameter (dkm)	meter (m)	decimeter (dm)	centimeter (cm)	millimeter (mm)
_____ meters	_____ meters	_____ meters	1 meter	_____ meter	_____ meter	_____ meter

Write the metric units in order from *smallest* to *largest*.

2. _____meter, _____meter, _____meter, meter, _____meter, _____meter, _____meter

Write the measurement of each line segment in millimeters. Write the measurement in centimeters.

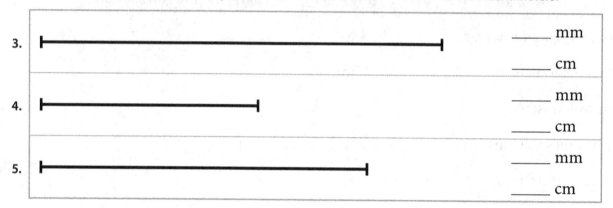

3. _____ mm _____ cm

4. _____ mm _____ cm

5. _____ mm _____ cm

Shade the Celsius temperature.

6. 20° 10° 12°C   **7.** 0° -10° -2°C   **8.** 0° -10° -8°C

Match.

_____ **9.** normal body temperature   **A** 100°C

_____ **10.** boiling point of water   **B** 37°C

_____ **11.** freezing point of water   **C** 0°C

Write >, <, or = to make the statement true.

12. 2×53 ◯ $(2 \times 50) + (2 \times 3)$

13. $a + b$ ◯ $b + a$

14. $(3 + 5) + 9$ ◯ $3 + (5 + 9)$

15. $\frac{25}{5}$ ◯ $20 \div 5$

16. 0.9 ◯ $\frac{1}{2}$

17. $21 + 2.3$ ◯ $2.1 + 23$

Write the value of *n* for the part-part-whole model.

18.

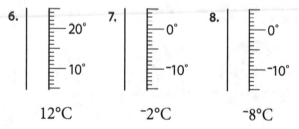

12			
n	n	n	n
n =			

19.

50	
25	n
n =	

Equivalent Ratios

Name _____

> **Ratios** can be used to show how two quantities are related. **Equivalent ratios** describe the same relationship in higher terms or lower terms.

Write an equivalent ratio in higher terms.

1. $\dfrac{2}{5} = \dfrac{4}{10}$

2. $\dfrac{6}{4} =$

3. $3:9 =$ _____

4. 8 to $2 =$ _____

5. $\dfrac{7}{3} =$

6. $\dfrac{2}{7} =$

7. $4:6 =$ _____

8. 9 to $5 =$ _____

Write an equivalent ratio in lower terms.

9. $\div 2$

$\dfrac{8}{6} = \dfrac{4}{3}$

$\div 2$

10. $\div 4$

$\dfrac{8}{12} = \dfrac{}{3}$

$\div 4$

11. $\div$

$\dfrac{12}{9} =$

$\div$

12. $\div$

$\dfrac{10}{15} =$

$\div$

13. $3:9 =$ ___1:3___

$\div 3$

14. $4:10 =$ _____

$\div$

15. $20:15 =$ _____

$\div$

16. $12:18 =$ _____

$\div$

17. 21 to $14 =$ _____

$\div$

18. 16 to $24 =$ _____

$\div$

19. 18 to $9 =$ _____

$\div$

20. 30 to $10 =$ _____

$\div$

> A **proportion** is an equation showing equivalent ratios.

Complete the proportion.

21. $\dfrac{3}{4} = \dfrac{}{8}$

22. $\dfrac{5}{3} = \dfrac{25}{}$

23. $\dfrac{4}{10} = \dfrac{}{100}$

24. $\dfrac{1}{7} = \dfrac{3}{}$

Complete the table. Answer the question.

25. The pattern for Mary's quilt calls for 3 red squares for every 4 white squares. Complete the ratio table to find how many white squares she will need for 15 red squares.

red	3	6			15
white	4				

26. What do ratios show?

_____

_____

27. What is a proportion?

_____

_____

> **Capacity** is the amount a container will hold.

Mark the better unit for capacity.

1. | one teaspoon of medicine |

   ○ mL   ○ L

2. | a fish aquarium |

   ○ mL   ○ L

3. | a cup of tea |

   ○ mL   ○ L

Complete the table.

4.

1 L	3 L	L
1,000 mL	mL	5,000 mL

Rename the units of capacity.

5. 7 L = _____ mL

6. 8,000 mL = _____ L

7. 2,500 mL = _____ L

8. 9.492 L = _____ mL

9. 600 mL = _____ L

10. 5.2 L = _____ mL

> **Mass**, or the amount of material in a substance, is measured in grams or kilograms.

Circle the items that would be measured with the given unit of mass.

11. | grams | raisin | large dog | paper clip | butterfly |

12. | kilograms | jellybean | horse | car | bed |

Complete the table.

13.

1 kg	4 kg	6 kg
g	g	g

Rename the units of mass.

14. 12,000 g = _____ kg

15. 4 kg = _____ g

16. 2.350 kg = _____ g

17. 5.2 kg = _____ g

18. 6,800 g = _____ kg

19. 0.225 kg = _____ g

Add or subtract. Rename if needed.

20.
$$\begin{array}{r} 8 \text{ m } 32 \text{ cm} \\ + 2 \text{ m } 42 \text{ cm} \\ \hline \end{array}$$

21.
$$\begin{array}{r} 9 \text{ m } 82 \text{ cm} \\ + 4 \text{ m } 51 \text{ cm} \\ \hline \end{array}$$

22.
$$\begin{array}{r} 3 \text{ cm } 5 \text{ mm} \\ - 2 \text{ cm } 9 \text{ mm} \\ \hline \end{array}$$

23.
$$\begin{array}{r} 4 \text{ L } 325 \text{ mL} \\ - 2 \text{ L } 425 \text{ mL} \\ \hline \end{array}$$

Add or subtract. Write the answer in lowest terms.

24. $\frac{3}{4} + \frac{1}{4} + \frac{2}{4} =$

25. $\frac{5}{12} - \frac{2}{12} =$

26. $\frac{1}{3} + \frac{1}{3} + \frac{1}{3} =$

27.
$$\begin{array}{r} 6\frac{1}{4} \\ + 5\frac{3}{8} \\ \hline \end{array}$$

28.
$$\begin{array}{r} 4\frac{2}{3} \\ - 1\frac{2}{5} \\ \hline \end{array}$$

29.
$$\begin{array}{r} 4 \\ - 1\frac{3}{7} \\ \hline \end{array}$$

30.
$$\begin{array}{r} 7\frac{1}{3} \\ + 2\frac{1}{4} \\ \hline \end{array}$$

Map Scales

Name _____

Ratios can be used to compare the size of a map and the size of the actual object represented by the map. This map was made with a scale of 1 in. to 300 mi.

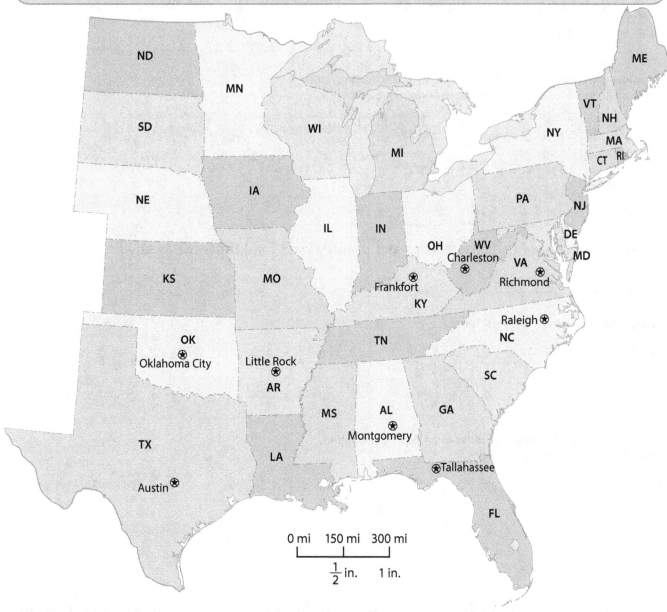

0 mi   150 mi   300 mi

$\frac{1}{2}$ in.   1 in.

Use the map and a ruler to complete the table.

	Cities	Approximate Distance
1.	Oklahoma City, OK, to Little Rock, AR	
2.	Richmond, VA, to Raleigh, NC	
3.	Little Rock, AR, to Tallahassee, FL	
4.	Frankfort, KY, to Charleston, WV	
5.	Austin, TX, to Montgomery, AL	

Write the ratio of fruit crystals to water in three ways.

> Shawna needs 2 cups of fruit crystals for every 6 cups of water to make punch for a party.

1.

Ratio	Word Form	Ratio Form	Fraction Form
crystals to water			

Complete the table to find out how much water and fruit crystals Shawna will need for 2, 3, and 4 times the original recipe.

2.

Increasing the Recipe				
	× 1	× 2	× 3	× 4
crystals	2			
cups of water	6			

3. Write the ratio of fruit crystals to water in three ways for 4 times the original recipe.

_____ _____ _____

Solve. Write the answer in lowest terms.

4. $\frac{2}{3} \times \frac{3}{4} =$

5. $\frac{1}{5} \times \frac{1}{2} =$

6. $\frac{3}{8} \times \frac{2}{7} =$

7. $\frac{3}{4} \times \frac{5}{6} =$

8. $\frac{5}{9} \times \frac{1}{2} =$

9. $\frac{3}{8} \times \frac{2}{3} =$

10. $\frac{1}{2} \times \frac{1}{2} =$

11. $\frac{2}{7} \times \frac{3}{5} =$

Solve by renaming the mixed number as an improper fraction. Write the answer in lowest terms.

12. $3 \times 2\frac{1}{4} =$

13. $2 \times 2\frac{2}{3} =$

14. $4\frac{1}{3} \times 1\frac{1}{2} =$

15. $2\frac{2}{5} \times 1\frac{1}{4} =$

Complete the table.

16.

Feet	Inches
2	
4	
6	

17.

Quarts	Gallons
4	
12	
24	

18.

Tons	Pounds
1	
3	
6	

19.

Days	Hours
1	
3	
6	

Complete the fact.

20. 1 mile = _____ feet

21. 3 feet = _____ yard

22. 1 pound = _____ ounces

23. 120 seconds = _____ minutes

Rates

Use the unit rate to solve. Label your answer.

1. In the meat department, ribs were on sale for $3.95 per pound. What was the cost for 5 pounds?

   > $3.95 : 1 lb; $19.75

2. Dad drove an average speed of 60 miles per hour. How many miles did he travel in 5 hours?

3. Jared's car traveled 21 miles on 1 gallon of gas. How many miles could he travel on 5 gallons?

Solve to find the unit rate. Label your answer.

4. Samuel earned $75 in 5 weeks mowing lawns. How much did he earn each week?

5. Craig scored 42 points in 6 games. What was the average number of points he scored per game?

6. There are 48 ounces of pop in 3 bottles. How many ounces are in each bottle?

Find the unit rate.

7. Marcos earned $50 in 5 hours.

   $$\frac{\$50}{5} = \$10/hr$$

8. Miriam typed 186 words in 3 minutes.

9. Five pounds of apples cost $3.75.

10. 8 projects in 4 days

11. $24 : 2 hr

12. 520 mi : 8 hr

13. 360 mi : 15 gal

14. $1,500 : 5 days

15. 48 points : 6 games

Find the distance traveled in the time given.

16. 3.5 hours at 60 mph = _____ miles

17. 9 hours at 70 mph = _____ miles

18. 5 days at 225 mi/day = _____ miles

19. 20 seconds at 12 ft/sec = _____ feet

20. 0.5 hour at 10 km/hr = _____ kilometers

21. 9 hours at 64 mph = _____ miles

22. 1.5 hours at 60 mph = _____ miles

Write an equivalent ratio in higher terms and in lower terms.

1. $\frac{2}{4} = \frac{4}{8}$

$= \frac{1}{2}$

2. $\frac{3}{9} = $ _____

$= $ _____

3. 12 to 6 = _____

$= $ _____

4. 5 : 10 = _____

$= $ _____

Complete the proportion.

5. $\frac{3}{4} = \frac{}{8}$

6. $\frac{4}{8} = \frac{}{24}$

7. $\frac{10}{20} = \frac{1}{}$

8. $\frac{9}{6} = \frac{}{2}$

9. $\frac{1}{2} = \frac{}{18}$

10. $\frac{15}{5} = \frac{}{1}$

11. $\frac{5}{8} = \frac{20}{}$

12. $\frac{28}{7} = \frac{}{1}$

Solve. Rename if necessary.

13. 4 hr 15 min
   + 3 hr 50 min

14. 5 tn 320 lb
   − 3 tn 500 lb

15. 4 yd 20 in.
   + 2 yd 30 in.

16. 6 gal
   − 2 gal 1 qt

17. 6 lb 4 oz
   + 2 lb 5 oz

18. 8 pt 3 c
   − 4 pt 2 c

19. 3 ft 5 in.
   − 2 ft 3 in.

20. 3 gal 3 qt
   + 5 gal 4 qt

Solve.

21. $40)\overline{429}$

22. $50)\overline{450}$

23. $22)\overline{1,342}$

24. $14)\overline{1,848}$

25. $42)\overline{7,432}$

26. $25)\overline{1,200}$

27. $20)\overline{\$75.00}$

28. $12)\overline{\$69.36}$

Complete the table.

29.

cm	mm
1	
2	
3	

30.

km	m
1	
3	
6	

31.

L	mL
1	
2	
4	

32.

kg	g
1	
4	
8	

Ratios & Percentages

Write the ratio as a percentage.

1. $\frac{30}{100} =$ _____

2. $\frac{25}{100} =$ _____

3. $59 : 100 =$ _____

4. 99 to $100 =$ _____

Write the percentage in fraction form with 100 as the denominator.

5. $16\% =$ _____

6. $25\% =$ _____

7. $3\% =$ _____

8. $48\% =$ _____

Complete the table by writing the equivalent fraction or percentage.

One hundred boys and girls were surveyed about their favorite foods.

9.

Food	Ratio	Percentage
pizza	$\frac{40}{100}$	
macaroni and cheese		30%
corn dogs	$\frac{25}{100}$	
other		5%

Write the percentage as a ratio in fraction form in lowest terms.

10. $20\% = \frac{20}{100} = \frac{1}{5}$

11. $10\% =$ _____

12. $50\% =$ _____

13. $80\% =$ _____

14. $25\% =$ _____

15. $75\% =$ _____

16. $15\% =$ _____

17. $35\% =$ _____

Complete the proportion. Write the ratio as a percentage.

18. $\frac{3}{10} = \frac{}{100} =$ _____

19. $\frac{5}{25} = \frac{}{100} =$ _____

20. $\frac{3}{5} = \frac{}{100} =$ _____

21. $\frac{3}{50} = \frac{}{100} =$ _____

Follow the steps to complete the table.

Fifty students were surveyed about their pets.

	Pet	Students	Ratio	$\frac{?}{100}$	Percentage
22.	dog	30	$\frac{30}{50}$		
23.	cat	10			
24.	hamster	5			
25.	fish	5			

Match.

_____ 26. ratio

_____ 27. proportion

_____ 28. equivalent ratio

_____ 29. percentage

A describes the same relationship in higher or lower terms

B a comparison of 2 quantities

C a ratio comparing the quantity to 100

D an equation showing equivalent ratios

Use the scale to complete the table.

> Anya is making a scale drawing of her garden.

Scale: 1 cm = 5 m

1.

	Scale Drawing		Actual Garden	
	Width	Length	Width	Length
garden	3 cm	4 cm	15 m	20 m
tomatoes	1 cm	2 cm		
squash	1 cm	3 cm		
lettuce	1 cm	1 cm		
corn	2 cm	3 cm		

Solve. Write the answer in lowest terms.

2. $4 \times \frac{2}{3} =$

3. $2 \times \frac{5}{6} =$

4. $1 \times \frac{7}{8} =$

5. $10 \times \frac{3}{5} =$

6. $\frac{1}{2} \times 3 =$

7. $\frac{3}{4} \times 5 =$

8. $\frac{2}{7} \times 9 =$

9. $\frac{1}{9} \times 8 =$

10. $\frac{2}{3} \times \frac{1}{2} =$

11. $\frac{1}{4} \times \frac{2}{5} =$

12. $\frac{5}{6} \times \frac{3}{4} =$

13. $\frac{3}{7} \times \frac{3}{5} =$

14. $3 \times \frac{4}{5} =$

15. $\frac{2}{3} \times 8 =$

16. $\frac{2}{7} \times \frac{9}{10} =$

17. $\frac{8}{9} \times \frac{2}{4} =$

Plot the mixed number on the number line. Round to the nearest whole number.

18. $1\frac{3}{4}$ ⬭

19. $3\frac{5}{8}$ ⬭

20. $9\frac{4}{10}$ ⬭

21. $5\frac{4}{6}$ ⬭

22. $\frac{1}{3}$ ⬭

23. $\frac{2}{5}$ ⬭

24. $\frac{4}{7}$ ⬭
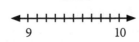

25. $\frac{4}{9}$ ⬭
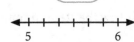

Write the equivalent unit of time.

26. 1 minute = _____ seconds

27. 1 decade = _____ years

28. 1 hour = _____ minutes

29. 1 century = _____ years

30. 1 day = _____ hours

31. 1 millennium = _____ years

32. 1 year = _____ days

33. 1 month = _____ days

Decimals & Percentages

Name _____

Shade the grid to represent the decimal. Write the decimal in fraction form and percentage form.

1.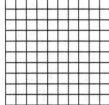

$0.25 = \dfrac{}{100} =$ _____

2.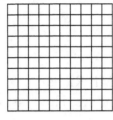

$0.50 = \dfrac{}{100} =$ _____

3.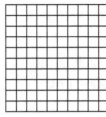

$0.75 = \dfrac{}{100} =$ _____

4.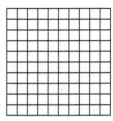

$0.29 = \dfrac{}{100} =$ _____

Complete the chart.

5.

Fraction	Percentage	Decimal
$\dfrac{32}{100}$		
$\dfrac{4}{25}$		
$\dfrac{10}{50}$		
$\dfrac{9}{20}$		
$\dfrac{7}{10}$		

6.

Decimal	Percentage	$\dfrac{?}{100}$	Fraction in Lowest Terms
0.10			
0.25			
0.40			
0.60			
0.75			

Write >, <, or = to compare.

7. $\dfrac{1}{4}$ ◯ 30%

8. $\dfrac{1}{2}$ ◯ 49%

9. $\dfrac{3}{4}$ ◯ 92%

10. 10% ◯ $\dfrac{1}{10}$

11. 25% ◯ $\dfrac{1}{2}$

12. 75% ◯ $\dfrac{3}{4}$

13. 0.42 ◯ 50%

14. 0.85 ◯ 85%

15. 0.7 ◯ 7%

Fill in the blanks with the correct percentage.

16.

$\dfrac{4}{4} = \underline{100}\%$

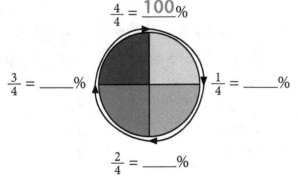

$\dfrac{3}{4} =$ _____ %

$\dfrac{1}{4} =$ _____ %

$\dfrac{2}{4} =$ _____ %

Solve. Label your answer.

17. Bailey answered 42 out of 50 questions correctly on the science test. What percentage of his answers were correct?

18. On his math test, Zachary answered 23 out of 25 problems correctly. What percentage of his answers were correct?

Use the unit rate to solve. Label your answer.

1. Madaline earns $9.00 per hour. How much will she earn in 10 hours?

2. Samuel scored 8 points in each game. How many games did he play if he scored a total of 32 points?

Solve to find the unit rate. Label your answer.

3. The Moore family traveled 750 miles on vacation on 30 gallons of gasoline. What is the mileage rate of the car?

4. Mr. Richards traveled 120 miles in 2 hours. How fast did he travel per hour?

Write *acute*, *right*, or *obtuse* to classify the triangle by the measures of its angles.

5.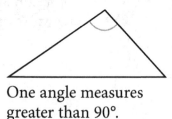
One angle measures greater than 90°.

_____

6.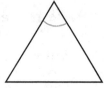
All angles measure less than 90°.

_____

7.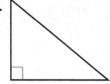
One angle measures 90°.

_____

Write *equilateral*, *isosceles*, or *scalene* to classify the triangle by the lengths of its sides.

8.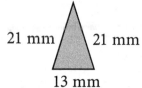
21 mm   21 mm
13 mm
Two sides are equal.

_____

9.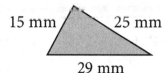
15 mm   25 mm
29 mm
No sides are equal.

_____

10.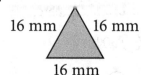
16 mm   16 mm
16 mm
All sides are equal.

_____

Complete the table.

11.

Rule: × 4	
Input	Output
8	
80	
800	
8,000	

12.

Rule: ÷ 5	
Input	Output
25	
250	
2,500	
25,000	

13.

Rule: ÷ 8	
Input	Output
24	
320	
4,000	
48,000	

14.

Rule: × 9	
Input	Output
30	
50	
600	
700	

Percentage of a Number

Name _____

Complete the proportion to find the percentage of the number.

1. 10% of $50

$$\frac{10}{100} = \frac{}{\$50}$$

2. 20% of $20

$$\frac{20}{100} = \frac{}{\$20}$$

3. 40% of $25

$$\frac{40}{100} = \frac{}{\$25}$$

Use mental math to find the percentage of the number.

4. 10% of 20 = _____

   20% of 20 = _____

   30% of 20 = _____

5. 10% of 40 = _____

   30% of 40 = _____

   50% of 40 = _____

6. 10% of 50 = _____

   40% of 50 = _____

   60% of 50 = _____

Multiply by the decimal to find the percentage of the number.

7. 10% of 30 = __3__

   (__0.10__ × __30__ =

   __3.00__)

8. 30% of 20 = _____

   (_____ × _____ =

   _____)

9. 50% of 30 = _____

   (_____ × _____ =

   _____)

10. 10% of 40 = _____

11. 30% of 40 = _____

12. 50% of 50 = _____

13. 10% of 5 = _____

14. 30% of 60 = _____

15. 50% of 70 = _____

Find the discount and the final cost.

16.

Price	Discount		Final Cost
$40	– (10%)	$4	=
$40	– (30%)		=
$60	– (50%)		=
$60	– (75%)		=

Solve. Label your answer.

17. Brodie earned $120 mowing lawns. He will give 10% to the church on Sunday. How much will he give?

18. Brodie put 50% of the $120 in savings. How much did he save?

19. Taryn's soccer team won 90% of the 30 games they played. How many games did they win?

20. At the baseball game, 80% of the 500 seats were filled. How many people attended the game?

Write the ratio as a percentage.

1. $\frac{30}{100} =$ _____

2. $\frac{55}{100} =$ _____

3. 45 to 100 = _____

4. $90 : 100 =$ _____

5. $\frac{8}{100} =$ _____

6. 100 to 100 = _____

Write the percentage as a ratio in fraction form in lowest terms.

7. $50\% = \frac{50}{100} =$

8. $25\% = \frac{25}{100} =$

9. $75\% =$

10. $10\% =$

11. $60\% =$

12. $35\% =$

Use the table to make a circle graph showing the favorite pies.

13.

Favorite Pies	
chocolate	$\frac{1}{2}$
blueberry	$\frac{1}{4}$
apple	$\frac{1}{4}$

Write the factors of the number. Label the number *prime* or *composite*.

14. 2: _____

_____

15. 4: _____

_____

16. 6: _____

_____

17. 7: _____

_____

18. 8: _____

_____

19. 10: _____

_____

Match.

_____ 20. linear   **A** liter

_____ 21. capacity   **B** meter

_____ 22. mass   **C** gram

Complete the fact.

23. 1 km = _____ m

24. 1 m = _____ cm = _____ mm

25. 1 cm = _____ mm

Probability

Use the spinner to find the probability of the event. Write the answer as a fraction. Simplify.

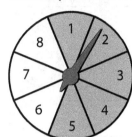

1. P(shaded) = $\frac{5}{8}$

2. P(unshaded) = _____

3. P(odd) = _____

4. P(even and shaded) = _____

5. P(odd and shaded) = _____

6. P(odd and unshaded) = _____

Bag of marbles
1 yellow
2 green
3 red
4 blue

Choose one marble at random from the bag.
Write the probability of the event as a fraction and a percentage.

7. P(yellow) = $\frac{1}{10}$ = _____

8. P(green) = ☐ = _____

9. P(red) = ☐ = _____

10. P(blue) = ☐ = _____

11. P(black) = ☐ = _____

12. P(not blue) = ☐ = _____

13. P(green and blue) = ☐ = _____

14. P(not black) = ☐ = _____

Complete the table by writing the frequency.
Write the probability of the event as a fraction and a percentage.

Macy rolled a number cube 25 times. She tallied how many times the cube landed showing each number.

Number Cube Rolls: 25 Times		
Number	Tally	Frequency
1					
2	̶T̶H̶L̶	
3					
4	̶T̶H̶L̶ ||	
5						
6					

15. P(1) = $\frac{3}{25}$ = **12%**

16. P(2) = ☐ = _____

17. P(3) = ☐ = _____

18. P(4) = ☐ = _____

19. P(5) = ☐ = _____

20. P(6) = ☐ = _____

21. P(even) = ☐ = _____

22. P(odd) = ☐ = _____

Find the percentage.

23. The probability of rain on Wednesday is 4 out of 5. _____

24. The probability of snow on Saturday is 6 out of 10. _____

Write the fraction as a decimal and as a percentage.

1. $\frac{3}{20}$ = _____ = _____

2. $\frac{3}{5}$ = _____ = _____

3. $\frac{1}{2}$ = _____ = _____

4. $\frac{4}{10}$ = _____ = _____

5. $\frac{2}{25}$ = _____ = _____

6. $\frac{3}{4}$ = _____ = _____

Write the percentage in fraction form in lowest terms.

7. 10% = $\frac{10}{100}$ = _____

8. 25% = $\frac{25}{100}$ = _____

9. 50% = _____

10. 75% = _____

11. 20% = _____

12. 80% = _____

Write the number in expanded form.

13. 0.245 = _____

14. 4.012 = _____

15. 9.307 = _____

16. 5.029 = _____

17. 7.968 = _____

Complete the table.

	Standard form	324,029,716
18.	Word form	
19.	Expanded form	
20.	Expanded form with multiplication	

Write >, <, or = to compare.

21. 12 ft ◯ 5 yd

22. 20 oz ◯ 2 c

23. 1 lb ◯ 16 oz

24. 40 in. ◯ 4 ft

25. 3 pt ◯ 7 c

26. 3,200 lb ◯ 3 tn

27. 2 mi ◯ 8,400 ft

28. 8 qt ◯ 2 gal

29. 8 pt ◯ 2 c

Ratios, Proportions & Percentages Review

Name _____

Mark the correct answer.

1. Mark the ratio that expresses 4 to 7.
   - ○ 28
   - ○ 4 : 7
   - ○ 4 × 7
   - ○ 4 ÷ 7

2. If a recipe takes 1 cup of sugar to 3 cups of flour, how many cups of flour will be needed with 3 cups of sugar?
   - ○ 6 c
   - ○ 7 c
   - ○ 9 c
   - ○ 12 c

3. Mark the unit rate for $180 : 15 hours.
   - ○ $5/hr
   - ○ $18/hr
   - ○ $9/hr
   - ○ $12/hr

4. Mark the distance traveled in 3.5 hours at 60 miles per hour.
   - ○ 180 mi
   - ○ 210 mi
   - ○ 300 mi
   - ○ 350 mi

5. Mark the answer that completes the proportion: $\frac{9}{1} = \frac{?}{3}$.
   - ○ 27
   - ○ 18
   - ○ 30
   - ○ 36

6. Mark the correct proportion.
   - ○ $\frac{3}{4} = \frac{80}{100}$
   - ○ $\frac{7}{8} = \frac{21}{24}$
   - ○ $\frac{1}{2} = \frac{20}{50}$
   - ○ $\frac{63}{25} = \frac{9}{5}$

7. Mrs. Pace paid $2.95 a pound for ribs. How much did she pay for 3 pounds?
   - ○ $7.85
   - ○ $6.25
   - ○ $8.85
   - ○ $9.00

8. Mark the answer that completes the proportion for these similar figures.

 1 in.
3 in.

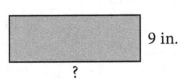 9 in.
?

   - ○ $\frac{1}{9} = \frac{1}{9}$
   - ○ $\frac{1}{9} = \frac{3}{27}$
   - ○ $\frac{1}{9} = \frac{9}{18}$
   - ○ $\frac{1}{9} = \frac{3}{18}$

9. On the field trip there were 10 students for every adult. If there were 50 students on the trip, how many adults were there?
   - ○ 4
   - ○ 6
   - ○ 5
   - ○ 7

10. The Satterfields traveled 560 miles by train in 8 hours. How fast did the train travel per hour?
    - ○ 60 mph
    - ○ 80 mph
    - ○ 70 mph
    - ○ 90 mph

11. Juan completed 50 math facts in one minute. At this rate, how many facts could he complete in 3 minutes?
    - ○ 100
    - ○ 150
    - ○ 125
    - ○ 175

Mark the correct proportion.

12. ○ $\frac{75}{25} = \frac{3}{1}$ ○ $\frac{4}{15} = \frac{24}{85}$

    ○ $\frac{12}{5} = \frac{48}{25}$ ○ $\frac{6}{8} = \frac{13}{16}$

13. ○ $\frac{7}{13} = \frac{21}{26}$ ○ $\frac{4}{8} = \frac{16}{24}$

    ○ $\frac{8}{140} = \frac{16}{280}$ ○ $\frac{4}{5} = \frac{12}{20}$

Mark the correct percentage.

14. $20 : 100 = \underline{\ ?\ }$

- ⭘ 20% ⭘ 40%
- ⭘ 30% ⭘ 50%

15. $5 \text{ to } 10 = \underline{\ ?\ }$

- ⭘ 5%   ⭘ 10%
- ⭘ 50% ⭘ 15%

16. $\frac{45}{100} = \underline{\ ?\ }$

- ⭘ 100% ⭘ 45%
- ⭘ 40%   ⭘ 50%

17. $\frac{3}{10} = \underline{\ ?\ }$

- ⭘ 3%   ⭘ 30%
- ⭘ 10% ⭘ 40%

18. $\frac{20}{50} = \underline{\ ?\ }$

- ⭘ 70% ⭘ 30%
- ⭘ 20% ⭘ 40%

19. $\frac{3}{4} = \underline{\ ?\ }$

- ⭘ 25% ⭘ 50%
- ⭘ 75% ⭘ 80%

20. Mark the unit rate for $72 : 9$ hours.

- ⭘ $8/hr   ⭘ $14/hr
- ⭘ $9/hr   ⭘ $18/hr

Mark the fraction form in lowest terms.

21.	60%	⭘ $\frac{6}{10}$	⭘ $\frac{3}{5}$	⭘ $\frac{3}{10}$
22.	25%	⭘ $\frac{25}{100}$	⭘ $\frac{5}{10}$	⭘ $\frac{1}{4}$
23.	40%	⭘ $\frac{2}{5}$	⭘ $\frac{4}{5}$	⭘ $\frac{1}{5}$

Match the correct decimal form.

_____ 24. 5%   **A** 0.52

_____ 25. 35%   **B** 0.05

_____ 26. 52%   **C** 0.35

Mark the correct percentage.

27.	0.7	⭘ 70%	⭘ 7%	⭘ 17%
28.	0.14	⭘ 14%	⭘ 41%	⭘ 40%
29.	0.25	⭘ 50%	⭘ 20%	⭘ 25%

Find the percentage of the number.

30. 10% of 40 = _____

31. 20% of 30 = _____

32. 30% of 90 = _____

33. 40% of 50 = _____

Name _____

Mark the answer.

1. What is the measure of the missing angle?

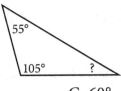

A. 20°   C. 60°

B. 40°   D. 80°

2. What is the measure of the missing angle?

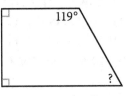

A. 21°   C. 50°

B. 39°   D. 61°

3. What is the volume of the cube?

4 ft

A. 12 ft³   C. 32 ft³

B. 16 ft³   D. 64 ft³

4. Mark the figure that shows a line of symmetry.

A.   C.

B.   D.

5. Mark the shapes that are congruent.

A.    C.

B.   D.

6.
```
  15 ft 8 in.
+  8 ft 9 in.
```

A. 22 ft 1 in.   C. 24 ft 5 in.

B. 23 ft 5 in.   D. 25 ft 1 in.

7.
```
  1 yd 2 ft
+ 6 yd 2 ft
```

A. 7 yd 2 ft   C. 8 yd

B. 8 yd 1 ft   D. 9 yd 2 ft

8.
```
  16 ft  3 in.
−  4 ft 11 in.
```

A. 11 ft 4 in.   C. 12 ft

B. 11 ft 6 in.   D. 12 ft 8 in.

9.
```
  3 hr 16 min
+ 3 hr 45 min
```

A. 6 hr 50 min   C. 7 hr

B. 6 hr 59 min   D. 7 hr 1 min

10.
```
  5 hr 20 min
− 2 hr 55 min
```

A. 1 hr 35 min   C. 2 hr 25 min

B. 2 hr 15 min   D. 3 hr 35 min

Mark the answer.

11. Round 13.786 to the nearest tenth.
    A. 10   C. 13.79
    B. 14   D. 13.8

12. Round 378,496,214 to the nearest one million.
    A. 378,500,000   C. 388,000,000
    B. 378,000,000   D. 400,000,000

13. What is the value of 5 in 956,380.14?
    A. fifty thousand   C. fifty million
    B. five million   D. five hundred

14. seventeen million, forty-nine thousand, eight hundred thirty-three
    A. 1,749,833   C. 17,049,833
    B. 7,490,833   D. 17,490,813

15. equal or equivalent to four and six tenths
    A. $4\frac{6}{10}$ C. $4\frac{3}{5}$
    B. 4.6   D. all of the above

16. twelve and nine hundredths
    A. 12.09   C. $12\frac{90}{100}$
    B. 12.9   D. all of the above

17. Luis began raking the leaves at 9:20 a.m. He finished 45 minutes later. What time did he finish raking?
    A. 9:50 a.m.   C. 10:00 a.m.
    B. 9:55 a.m.   D. 10:05 a.m.

18. Which characteristic describes a cube?
    A. 6 square faces   C. 6 rectangular faces
    B. 4 square faces   D. 4 rectangular faces

19. The principal is replacing batteries in all emergency flashlights at school. Each flashlight takes 2 batteries. How many batteries are needed for the 17 flashlights?
    A. 30   C. 38
    B. 34   D. 42

20. A gallon of syrup cost $32.95 last year. This year the cost for a gallon of syrup is $34.00. What is the increase in cost?
    A. $1.05   C. $1.95
    B. $1.50   D. $2.05

21. The grocery store advertised a sale on cereal: 2 boxes for $7.28. What is the cost of 1 box of cereal?
    A. $3.49   C. $5.00
    B. $3.64   D. $4.32

Positive and Negative Numbers

Name _____

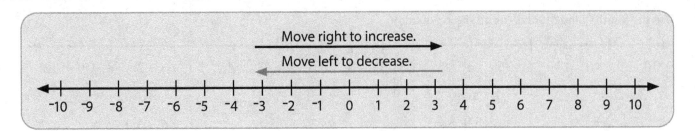

Use the number line to compare. Write > or <.

1. 3 ◯ ⁻3

2. ⁻2 ◯ 0

3. ⁻6 ◯ 4

4. ⁻10 ◯ 10

5. 0 ◯ ⁻5

6. 2 ◯ 0

7. ⁻7 ◯ 6

8. ⁻4 ◯ 4

Write the missing numbers.

9.
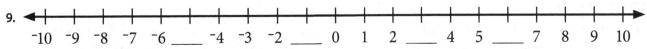
⁻10 ⁻9 ⁻8 ⁻7 ⁻6 ____ ⁻4 ⁻3 ⁻2 ____ 0 1 2 ____ 4 5 ____ 7 8 9 10

10.
⁻10 ⁻9 ____ ⁻7 ⁻6 ⁻5 ⁻4 ⁻3 ____ ⁻1 ____ 1 2 3 4 ____ 6 7 8 9 10

Write the numbers from *least* to *greatest*.

11. | 10 0 3 4 | ____ ____ ____ ____

12. | 0 ⁻5 ⁻3 ⁻10 | ____ ____ ____ ____

13. | 3 ⁻3 ⁻8 2 | ____ ____ ____ ____

14. | 10 ⁻10 ⁻4 6 | ____ ____ ____ ____

Use the number line to find the answer. Draw an arrow to show the subtrahend. Solve.

15.
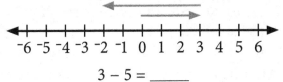
⁻6 ⁻5 ⁻4 ⁻3 ⁻2 ⁻1 0 1 2 3 4 5 6

$3 - 5 =$ _____

16.
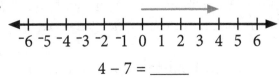
⁻6 ⁻5 ⁻4 ⁻3 ⁻2 ⁻1 0 1 2 3 4 5 6

$4 - 7 =$ _____

17.
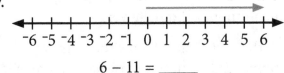
⁻6 ⁻5 ⁻4 ⁻3 ⁻2 ⁻1 0 1 2 3 4 5 6

$6 - 11 =$ _____

18.
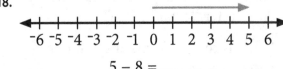
⁻6 ⁻5 ⁻4 ⁻3 ⁻2 ⁻1 0 1 2 3 4 5 6

$5 - 8 =$ _____

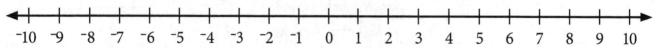

An addition equation can be used to join two positive or two negative integers.

Use the number line to solve the addition equation.

$$\xleftarrow{\hspace{0.5em}}\overset{\textstyle -10 \quad -9 \quad -8 \quad -7 \quad -6 \quad -5 \quad -4 \quad -3 \quad -2 \quad -1 \quad 0 \quad 1 \quad 2 \quad 3 \quad 4 \quad 5 \quad 6 \quad 7 \quad 8 \quad 9 \quad 10}{|\,|\,|\,|\,|\,|\,|\,|\,|\,|\,|\,|\,|\,|\,|\,|\,|\,|\,|\,|\,|}\xrightarrow{\hspace{0.5em}}$$

1. $2 + 3 = \underline{\quad 5 \quad}$ 2. $^-6 + ^-2 = \underline{\quad ^-8 \quad}$ 3. $^-1 + ^-1 = \underline{\quad\quad}$ 4. $4 + 2 = \underline{\quad\quad}$

5. $^-7 + ^-3 = \underline{\quad\quad}$ 6. $5 + 3 = \underline{\quad\quad}$ 7. $^-5 + ^-3 = \underline{\quad\quad}$ 8. $^-6 + ^-3 = \underline{\quad\quad}$

Write a positive or negative number to match the phrase.

9. lost five points $\underline{\quad ^-5 \quad}$

10. gained five pounds $\underline{\quad\quad}$

11. earned ten dollars $\underline{\quad\quad}$

12. two degrees below zero $\underline{\quad\quad}$

13. scored three points $\underline{\quad\quad}$

14. lost four buttons $\underline{\quad\quad}$

15. found nine pennies $\underline{\quad\quad}$

16. seventy degrees above zero $\underline{\quad\quad}$

Answer the question.

17. Aaron borrowed $5 from Dad and $10 from Mom. How much does he owe?

_____

18. Mia got $5 from each of 2 friends on her birthday. Her grandma gave her $10. How much does she have?

_____

19. The temperature was $^-2°$F at 7:00 p.m. It fell 5° during the night. What is the temperature now?

_____

20. The early morning temperature was $^-2°$F. By 8:00 a.m. the temperature had risen 5°. What is the temperature now?

_____

21. A shirt cost $10. A $3 discount is applied at checkout. Now what is the price of the shirt?

_____

22. The submarine was 5 fathoms below sea level. It dropped 2 more fathoms. What is the submarine's new elevation?

_____

Answer the question.

| Supplies for Josh's lemonade stand cost $15. Josh has $10. Mom will loan him the rest of the money he needs. | Day 1: He earned $5 and paid Mom $5. Day 2: He earned $10 and bought cups for $3. Day 3: He earned $8. |

23. How much money did Josh need to borrow from Mom? $\underline{\quad\quad}$

24. After he borrowed the money, how much money did he owe? $\underline{\quad\quad}$

25. How much money did he have left on day 1 after he paid Mom? $\underline{\quad\quad}$

26. After he bought more cups on day 2, how much money did he have? $\underline{\quad\quad}$

27. What is the total amount of money he had on day 3? $\underline{\quad\quad}$

28. At the end of day 3, how much more money did he have than when he started?

$\underline{\quad\quad}$

Adding Positive & Negative Numbers

Name _____

Using an Algebra Mat to Add Positive and Negative Numbers

1. Place the first addend. **8**

2. Place the second addend. **⁻5**

3. A positive counter and a negative counter cancel each other out to make 0. The answer is the number of counters remaining. **3**

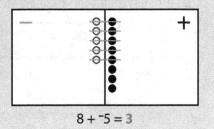

$$8 + {}^-5 = 3$$

The first addend is pictured on the mat. Draw counters for the second addend. Solve.

1.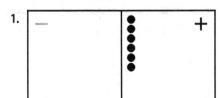

$$6 + {}^-4 = \underline{\quad 2 \quad}$$

2.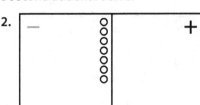

$${}^-7 + 4 = \underline{\qquad}$$

3.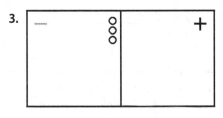

$${}^-3 + {}^-5 = \underline{\qquad}$$

4.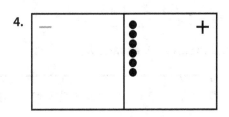

$$6 + 2 = \underline{\qquad}$$

Use an algebra mat and counters to find the sum.

5. $6 + 2 = \underline{\qquad}$ 6. ${}^-4 + {}^-2 = \underline{\qquad}$ 7. ${}^-6 + 3 = \underline{\qquad}$ 8. $9 + {}^-8 = \underline{\qquad}$

9. $4 + {}^-6 = \underline{\qquad}$ 10. $4 + 5 = \underline{\qquad}$ 11. $5 + {}^-5 = \underline{\qquad}$ 12. ${}^-3 + {}^-3 = \underline{\qquad}$

Write > or < to compare.

13. ${}^-4 \bigcirc {}^-7$ 14. ${}^-3 \bigcirc 3$ 15. $10 \bigcirc {}^-6$ 16. $5 \bigcirc {}^-7$

17. ${}^-8 \bigcirc {}^-1$ 18. $4 \bigcirc 5$ 19. ${}^-2 \bigcirc {}^-5$ 20. $4 \bigcirc {}^-4$

Write the numbers from *least* to *greatest*.

21. | ⁻5 | 0 | ⁻3 | 2 |

___ ___ ___ ___

22. | 1 | 4 | ⁻2 | ⁻6 |

___ ___ ___ ___

23. | ⁻5 | 5 | 2 | ⁻10 |

___ ___ ___ ___

24. | 4 | ⁻2 | 0 | ⁻6 |

___ ___ ___ ___

25. | ⁻8 | ⁻10 | 1 | 5 |

___ ___ ___ ___

26. | 1 | 12 | ⁻13 | 0 |

___ ___ ___ ___

Using a Number Line to Add Positive and Negative Numbers

Begin at 0. Draw an arrow to show the first addend.

To add a positive number, draw a second arrow that moves **right** from the first addend.

To add a negative number, draw a second arrow that moves **left** from the first addend.

$-8\,-7\,-6\,-5\,-4\,-3\,-2\,-1\ 0\ 1\ 2\ 3\ 4\ 5\ 6\ 7\ 8$

$-3 + 5 = 2$

$-8\,-7\,-6\,-5\,-4\,-3\,-2\,-1\ 0\ 1\ 2\ 3\ 4\ 5\ 6\ 7\ 8$

$-3 + {}^-5 = {}^-8$

Use the number line to find the answer. Draw an arrow to show the second addend. Solve.

1.
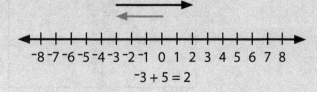

$-4 + 3 = \underline{\hphantom{00}}$

2.
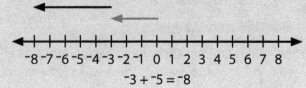

$5 + {}^-7 = \underline{\hphantom{00}}$

3.
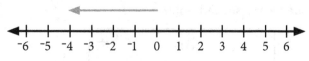

$5 + 1 = \underline{\hphantom{00}}$

4.
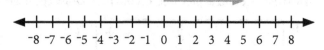

$-6 + {}^-2 = \underline{\hphantom{00}}$

Use the number line to solve the equation.

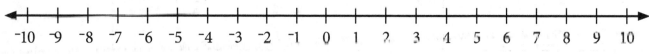

$-10\ \ -9\ \ -8\ \ -7\ \ -6\ \ -5\ \ -4\ \ -3\ \ -2\ \ -1\ \ \ 0\ \ \ 1\ \ \ 2\ \ \ 3\ \ \ 4\ \ \ 5\ \ \ 6\ \ \ 7\ \ \ 8\ \ \ 9\ \ \ 10$

5. $5 + {}^-5 = \underline{\hphantom{00}}$ **6.** $-4 + {}^-3 = \underline{\hphantom{00}}$ **7.** $3 + 6 = \underline{\hphantom{00}}$ **8.** $-9 + 4 = \underline{\hphantom{00}}$

Use the picture to answer the question.

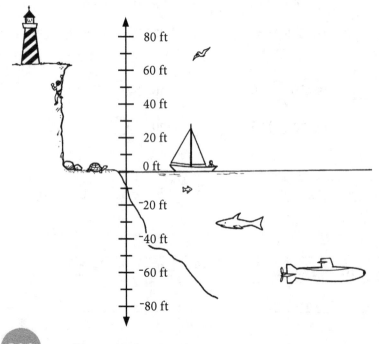

80 ft
60 ft
40 ft
20 ft
0 ft
-20 ft
-40 ft
-60 ft
-80 ft

9. Name an item that is at sea level.

_____

10. Write the positive elevation.

bird: _____70 ft_____

base of lighthouse: _____

foot of rock climber: _____

11. Write the negative elevation.

small fish: _____-10 ft_____

submarine: _____

shark: _____

Subtracting Negative Numbers

Name _____

Using an Algebra Mat to Subtract Positive and Negative Numbers

negative – negative

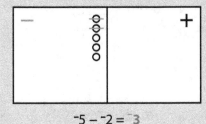

¯5 – ¯2 = ¯3

1. Place 5 negative counters.

2. Cross out (subtract) 2 negative counters.

positive – negative

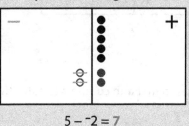

5 – ¯2 = 7

1. Place 5 positive counters.

2. Add pairs of +/– counters to get the needed 2 negative counters.

3. Cross out 2 negative counters.

Use an algebra mat and counters to solve the equation.

1. ¯3 – ¯2 = _____

2. 3 – ¯6 = _____

3. ¯4 – ¯1 = _____

4. 2 – ¯1 = _____

5. ¯2 – ¯1 = _____

6. 5 – ¯3 = _____

Draw arrows on the vertical number line to complete the sea otter's diving pattern. Solve. Label your answer.

7. The sea otter swam at a depth of 35 feet below the surface of the water. To catch a fish for lunch, he rose 15 feet. At what elevation was he swimming when he caught the fish?

   _____

8. After he caught the fish, he dove 30 feet to the sea floor. What was his new elevation?

   _____

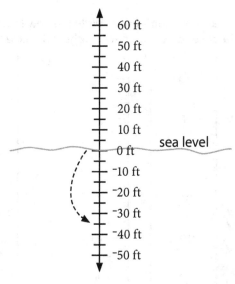

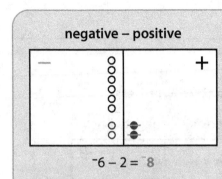

negative – positive

Add pairs of +/– counters to get the needed positive counters.

$^-6 - 2 = {}^-8$

Use an algebra mat and counters to solve the equation.

1. $^-4 - 1 =$ _____

2. $^-10 - 9 =$ _____

3. $^-3 - 3 =$ _____

4. $^-8 - 6 =$ _____

Use an algebra mat and counters to solve the equation.

5. $^-3 - {}^-2 =$ _____

6. $^-4 - 3 =$ _____

7. $^-5 - {}^-6 =$ _____

8. $6 - {}^-5 =$ _____

9. $7 - {}^-7 =$ _____

10. $^-8 - 3 =$ _____

Write the equation shown by the number line.

11.
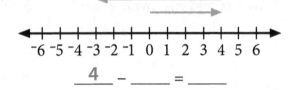

_____4_____ – _____ = _____

12.
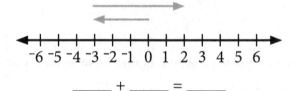

_____ + _____ = _____

13.
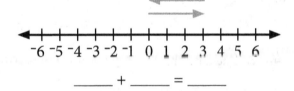

_____ + _____ = _____

14.
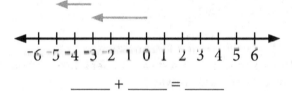

_____ + _____ = _____

Write the Celsius temperature. Write the new temperature. Shade the second thermometer to show the new temperature.

15.
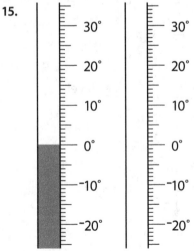

_____   _____
                 15° colder

16.
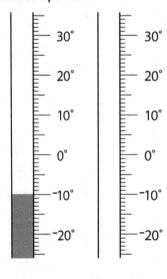

_____   _____
                 15° warmer

17.
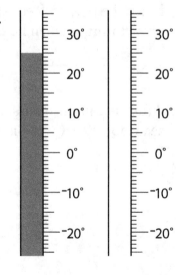

_____   _____
                 30° colder

Adding & Subtracting

Name _____

Using an Algebra Mat to Add

positive + negative	negative + positive	negative + negative

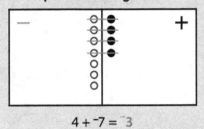 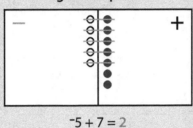 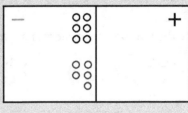

$4 + {}^-7 = {}^-3$ ${}^-5 + 7 = 2$ ${}^-6 + {}^-5 = {}^-11$

Use an algebra mat and counters to find the sum.

1. $5 + {}^-2 =$ _____
2. ${}^-3 + 4 =$ _____
3. ${}^-2 + {}^-6 =$ _____
4. ${}^-6 + 4 =$ _____

5. ${}^-4 + {}^-2 =$ _____
6. $8 + {}^-9 =$ _____
7. ${}^-3 + 8 =$ _____
8. ${}^-5 + 5 =$ _____

Using an Algebra Mat to Subtract

negative − negative	positive − negative	negative − positive

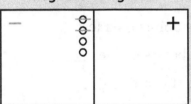 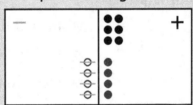 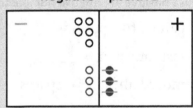

${}^-4 - {}^-2 = {}^-2$ $6 - {}^-4 = 10$ ${}^-5 - 3 = {}^-8$

Use an algebra mat and counters to find the difference.

9. ${}^-3 - {}^-1 =$ _____
10. $9 - {}^-6 =$ _____
11. ${}^-4 - {}^-5 =$ _____
12. ${}^-8 - {}^-2 =$ _____

13. ${}^-2 - {}^-5 =$ _____
14. $4 - {}^-5 =$ _____
15. ${}^-10 - 3 =$ _____
16. ${}^-3 - {}^-3 =$ _____

Use the number line to find the sum.

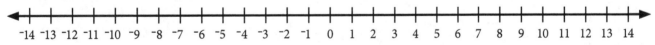

17. ${}^-6 + 9 =$ __3__
18. $3 + {}^-5 =$ _____
19. $4 + 6 =$ _____
20. ${}^-11 + 11 =$ _____

21. ${}^-4 + 12 =$ _____
22. ${}^-9 + {}^-3 =$ _____
23. $6 + 5 =$ _____
24. $1 + {}^-5 =$ _____

Circle the number that has the greater value.

1. ⁻2   ⁻5
2. 3   ⁻1
3. ⁻10   ⁻12
4. 0   ⁻5

Write > or < to compare.

5. ⁻4 ◯ ⁻2   6. 3 ◯ ⁻3   7. 5 ◯ ⁻10   8. ⁻9 ◯ 0

Write the numbers from *least* to *greatest*.

9. ⁻5   5   0   ⁻3
10. 4   1   ⁻5   6
11. ⁻3   0   3   ⁻2

____ ____ ____ ____   ____ ____ ____ ____   ____ ____ ____ ____

Use the number line to find the sum.

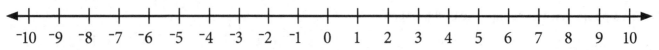

⁻10 ⁻9 ⁻8 ⁻7 ⁻6 ⁻5 ⁻4 ⁻3 ⁻2 ⁻1 0 1 2 3 4 5 6 7 8 9 10

12. 4 + ⁻8 = _____   13. 8 + ⁻1 = _____   14. ⁻5 + 3 = _____   15. ⁻3 + ⁻5 = _____

16. ⁻6 + ⁻3 = _____   17. ⁻9 + 6 = _____   18. ⁻2 + 4 = _____   19. 8 + ⁻3 = _____

20. ⁻6 + 6 = _____   21. 5 + ⁻10 = _____   22. 0 + ⁻6 = _____   23. ⁻2 + 6 = _____

Write a positive or negative number to match the phrase.

24. gained five pounds _____   25. missed three points on a test _____

26. lost four points _____   27. twenty feet above sea level _____

28. moved ahead two spaces _____   29. received two stickers _____

30. earned five dollars _____   31. owes ten dollars _____

Circle the numbers greater than 0. Draw a box around the numbers less than 0.

32.   4   ⁻10   ⁻2   1   12

Use the thermometer to solve.

50°
40°
30°
20°
10°
0°
⁻10°
⁻20°

33. In the morning the temperature was 10 degrees below 0. By 4:00 p.m. it was 45 degrees above 0. How many degrees did the temperature rise from morning to 4:00 p.m.?

_____

34. At 8:00 p.m. the temperature was 30 degrees lower than it was at 4:00 p.m. What was the temperature at 8:00 p.m.?

_____

Integers Review

Name _____

Mark the number with the least value.

1. 8   0   ⁻3
   ○   ○   ○

2. ⁻10   4   ⁻2
   ○   ○   ○

3. 1   ⁻9   4
   ○   ○   ○

4. ⁻3   ⁻1   ⁻8
   ○   ○   ○

Write > or < to compare.

5. 6 ◯ 10

6. ⁻7 ◯ 0

7. ⁻9 ◯ 3

8. ⁻4 ◯ ⁻6

Write the numbers from *least* to *greatest*.

9. | 4 | 8 | ⁻6 | ⁻10 |

____ ____ ____ ____

10. | ⁻3 | 3 | ⁻6 | 0 |

____ ____ ____ ____

11. | ⁻12 | 13 | ⁻10 | 11 |

____ ____ ____ ____

12. | ⁻1 | ⁻4 | 0 | ⁻10 |

____ ____ ____ ____

13. | ⁻5 | ⁻12 | ⁻13 | ⁻2 |

____ ____ ____ ____

14. | 0 | 5 | ⁻7 | 8 |

____ ____ ____ ____

Write the missing numbers on the number line. Answer the question.

15.
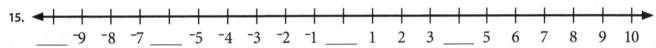
____ ⁻9 ⁻8 ⁻7 ____ ⁻5 ⁻4 ⁻3 ⁻2 ⁻1 ____ 1 2 3 ____ 5 6 7 8 9 10

16. In which direction does the value of the numbers increase?

_____

17. What happens as you move farther to the left of 0 on the number line?

_____

18. Explain why ⁻9 is less than ⁻4.

_____

19. What is the difference between ⁻5 and 3? _____

Use the number line to find the sum.

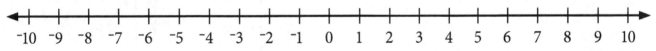
⁻10 ⁻9 ⁻8 ⁻7 ⁻6 ⁻5 ⁻4 ⁻3 ⁻2 ⁻1 0 1 2 3 4 5 6 7 8 9 10

20. 3 + ⁻6 = _____

21. ⁻4 + 5 = _____

22. ⁻9 + 9 = _____

23. 10 + ⁻4 = _____

24. 3 + ⁻5 = _____

25. 0 + ⁻8 = _____

26. 1 + ⁻5 = _____

27. ⁻7 + ⁻3 = _____

Write a positive or negative number to match the phrase.

28. jumped ahead four spaces _____

29. dove ten feet under water _____

30. lost five points _____

31. subtracted twenty dollars _____

The minuend counters are pictured on the mat. Draw counters for the subtrahend. Solve.

32.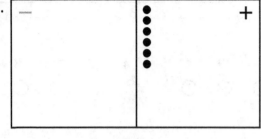

$6 - 10 =$ _____

33.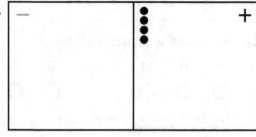

$4 - 8 =$ _____

34.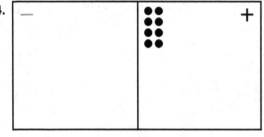

$8 - {}^-3 =$ _____

35.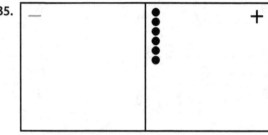

$6 - {}^-2 =$ _____

36.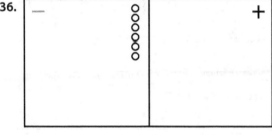

${}^-6 - 4 =$ _____

37.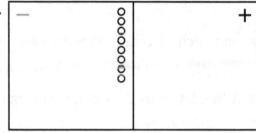

${}^-8 - 1 =$ _____

38.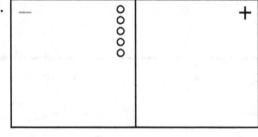

${}^-5 - {}^-3 =$ _____

39.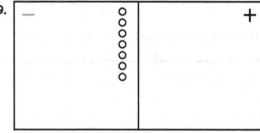

${}^-7 - {}^-4 =$ _____

Mark the answer.

1. the value of 7 in the number

    473,096

    A. 700
    B. 7,000
    C. 70,000
    D. 700,000

2. Round the decimal to the nearest tenth.

    3.861

    A. 4
    B. 3.8
    C. 3.9
    D. 3.86

3.

    A. angle A
    B. angle C
    C. angle ABC
    D. angle BAC

4.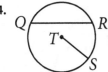

    A. circle QR
    B. circle T
    C. circle Q
    D. circle TS

5.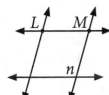

    A. plane LM
    B. plane n
    C. plane L
    D. plane M

6. $\overleftrightarrow{U \quad V}$

    A. line UV
    B. line U
    C. line V
    D. line VN

7. $\dfrac{3}{12} = \dfrac{?}{\underline{}}$

    A. $\dfrac{1}{2}$ C. $\dfrac{6}{12}$
    B. $\dfrac{1}{4}$ D. $\dfrac{3}{4}$

8. $\dfrac{7}{7} = \dfrac{?}{\underline{}}$

    A. $\dfrac{1}{7}$ C. 1
    B. 7   D. 0

9. $\dfrac{5}{9} - \dfrac{3}{9} = \dfrac{?}{\underline{}}$

    A. $\dfrac{8}{9}$ C. $\dfrac{1}{3}$
    B. $\dfrac{1}{9}$ D. $\dfrac{2}{9}$

10. $6 - \dfrac{2}{3} = \dfrac{?}{\underline{}}$

    A. $6\dfrac{3}{4}$ C. $\dfrac{4}{3}$
    B. $5\dfrac{1}{3}$ D. $6\dfrac{2}{3}$

11. $\dfrac{2}{5} \times \dfrac{2}{3} = \dfrac{?}{\underline{}}$

    A. $\dfrac{4}{8}$ C. $\dfrac{4}{15}$
    B. $\dfrac{1}{15}$ D. $\dfrac{2}{8}$

12. $\dfrac{1}{4} + \dfrac{2}{3} = \dfrac{?}{\underline{}}$

    A. $\dfrac{11}{12}$ C. $\dfrac{2}{5}$
    B. $\dfrac{1}{6}$ D. $\dfrac{1}{4}$

Mark the answer.

13.
$$72.39 + 7.93$$
A. 79.32
B. 80.32
C. 81.32
D. 82.32

14.
$$10,000 - 3,847$$
A. 6,047
B. 6,050
C. 6,153
D. 6,247

15.
$$203,781 + 159,236$$
A. 363,000
B. 363,017
C. 363,557
D. 364,027

16.
$$8.1 - 3.75$$
A. 4.35
B. 4.8
C. 5.35
D. 5.65

17.
$$\$12.25 - \$7.50$$
A. $4.25
B. $4.50
C. $4.75
D. $475

18.

Rule: + 12	
Input	Output
8	20
16	?
32	44

A. 24
B. 26
C. 28
D. 30

19.
$$671 \times 5$$
A. 3,005
B. 3,335
C. 3,355
D. 3,555

20.
$$866 \times 16$$
A. 13,102
B. 13,152
C. 13,208
D. 13,856

21.
$$29 \times 100 = \underline{\ ?\ }$$
A. 290
B. 2,000
C. 2,900
D. 29,000

22.
$$17 \times 14 = 14 \times \underline{\ ?\ }$$
A. 14
B. 15
C. 16
D. 17

23.
$$3\overline{)28}$$
A. 8
B. 9
C. 9 r1
D. 9 r7

24.
$$9\overline{)378}$$
A. 36
B. 38
C. 40
D. 42

25.
$$155 \div 5 = \underline{\ ?\ }$$
A. 11
B. 31
C. 35
D. 53

Line Plots & Stem-and-Leaf Plots

Name _____

Use the data to answer the question. Use the word bank to complete the statement.

1. What was the average game score?

   _____

   This is the _____ of the set of data.

2. What was the difference between the highest score and the lowest score? _____

   This is the _____ of the set of data.

3. What was the score that occured most often, or had the greatest frequency? _____

   This is the _____ of the set of data.

4. What was the middle value (score) of the data?

   _____

   This is the _____ of the set of data.

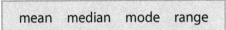

mean   median   mode   range

Carson recorded the last 5 game scores for his favorite basketball team: 71, 74, 88, 89, 88.

```
                                    X
       X     X                    X X
    +-+-+-+-+-+-+-+-+-+-+-+-+-+-+-+-+-+-+-+-+
    70        75        80        85        90
```
Game Scores

A **cluster** is a tight grouping of data on the line plot.
A **gap** is an empty space with no data on the line plot.

Use the line plot above to answer the question.

5. Where is there a cluster? _____

6. Between what scores is there the greatest gap?

   _____

A **stem-and-leaf plot** uses the tens digit of the data as its **stem** and the ones digit of the data as its **leaf**.

Use the stem-and-leaf plot to find the answer.

7. How many classes have 21 students? _____

8. How many classrooms are being counted? _____

9. What is the greatest number of students in a classroom? _____

10. What is the mean number of students in a classroom? _____

11. What is the median number of students in a classroom? _____

12. What number represents the mode of the data? _____

13. What is the range of the data? _____

Number of Students in Each Classroom	
Stem	Leaf
0	9
1	1 2 7 8 8
2	0 0 0 1 1 3 4

Key 2|1 = 21 students in the class

Solve. Label when applicable.

1. Peter's last five quizzes had the following scores: 12, 12, 14, 15, and 17. What was his average score?

2. Jarret and his family traveled 360 miles to Florida for vacation. If they averaged 60 miles per hour, how many hours did it take them to get to Florida?

3. Scott jumped rope for 5 minutes. He jumped 105 times the first minute, 130 the second, 115 the third, 135 the fourth, and 115 the fifth. What was his average number of jumps per minute?

4. Sun Valley Elementary School has 4 kindergarten classes. There are 20 students in section A, 22 in section B, 21 in section C, and 17 in section D. What is the average number of students per section?

Use the data to complete the stem-and-leaf plot.

Varsity basketball players' shoe sizes:
9, 9, 10, 10, 10, 10, 11, 12, 12, 13

5.

Players' Shoe Sizes	
Stem	Leaf
0	9

Key _____

Use the line plot to answer the question.

Josh recorded the number of red sports cars he saw each day for one week.

6. On which day did Josh see the most red sports cars? _____

7. On which day did he see only one? _____

8. On which days did Josh not see any? _____

9. How many did he see in all? _____

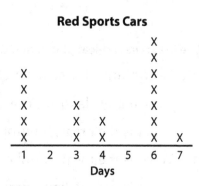

Red Sports Cars

Double Bar & Double Line Graphs

Name _____

Use the bar graph to find the answer.

1. Which grade has the most girl students?
   _____

2. Which grade has the fewest boy students?
   _____

3. Which grade has the greatest difference between boys and girls?

   _____

4. Which grade has the same number of boys and girls?

   _____

5. Find the range, mean, median, and mode for the number of girls in each grade.

   range: _____

   mean: _____

   median: _____

   mode: _____

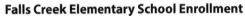

A **double bar graph** compares two sets of similar information on the same graph.

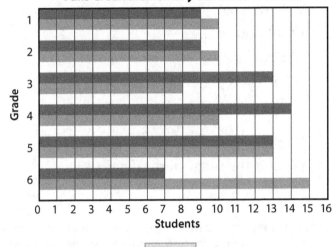

A **double line graph** compares how two amounts change over a period of time.

Use the line graph to find the answer.

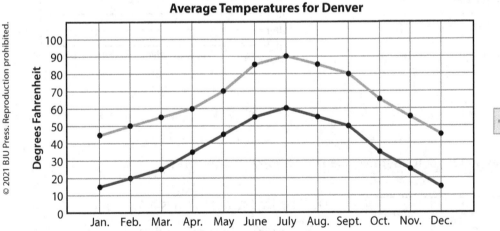

6. What is the average high temperature in May? _____

7. What is the average low temperature in January? _____

8. What is the range of the average high and low temperatures in March? _____

9. What temperature is the average high for April but also the average low for July? _____

10. What is the highest average temperature in Denver? _____

Answer the question. Use the data to complete the stem-and-leaf plot.

> Mr. and Mrs. Smith have 18 grandchildren.
>
> Grandchildren's ages: 1, 1, 3, 5, 6, 9, 10, 12, 12, 12, 14, 15, 17, 18, 19, 20, 20, 22

1. How many grandchildren are in each age group?

    0–9 years old: _____

    10–19 years old: _____

    20 or older: _____

2. What age represents the mode? _____

3. What is the mean age? _____

4. What is the median age? _____

5. What is the range of the data? _____

6.

Ages of Grandchildren	
Stem	Leaf

Key 2|0 = 20 years old

Use a calculator to find the range, mean, median, and mode of the data. Use the data from the chart to complete a single bar graph.

7. range: _____

8. mean: _____

9. median: _____

10. mode: _____

Colors of Cars at the Mall	
red	54
blue	62
silver	65
white	78
black	45

11.

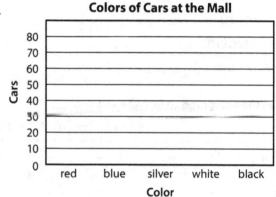

Colors of Cars at the Mall

Use the graph to mark the answer.

> Jordan recorded the first year's growth of his German Shepherd puppy on a line graph.

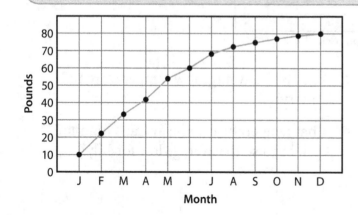

12. The line graph shows a
    ○ steady increase in growth.
    ○ steady decrease in growth.

13. The best title for the graph is
    ○ Man's Best Friend.
    ○ Puppy Food Increasing Growth.
    ○ Puppy Growth.

Pictographs & Circle Graphs

Name _____

Use the pictograph to find the answer.

Candy	Students
Favorite Kinds of Candy at Westwood Elementary School	
chocolate	🍬🍬🍬🍬🍬🍬🍬🍬🍬
caramel	🍬🍬🍬🍬
peppermint	🍬🍬🍬
fruit flavors	🍬🍬🍬🍬🍬🍬🍬
sour	🍬🍬🍬🍬🍬🍬🍬🍬🍬🍬🍬

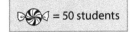 = 50 students

1. What type of candy is the favorite? _____

2. How many students like chocolate? _____

3. How many students like peppermint? _____

4. How many more students like fruit flavors than caramel? _____

Use the circle graph to find the answer.

5. Most of the dogs at the show were

_____.

6. The fewest dogs were

_____.

7. Were there more Cocker Spaniels or Labradors?

_____

8. What part of the graph represents Dachshunds?

    ○ $\frac{1}{4}$ ○ $\frac{1}{3}$ ○ $\frac{1}{2}$

Dogs at the Dog Show

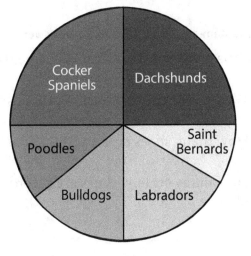

Find the range, mean, median, and mode for the data.

1.

20, 14, 18, 15, 18

range: _____

mean: _____

median: _____

mode: _____

2.

100, 95, 86, 94, 85

range: _____

mean: _____

median: _____

mode: _____

Use the line plot to answer the question.

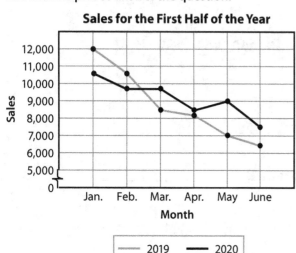

Sales for the First Half of the Year

3. Which year had a decrease in sales each month?

_____

4. Which month and year had the most sales?

_____

5. Which year had a smaller range of sales?

_____

Use the stem-and-leaf plot to find the answer.

6. What is the price of the most expensive athletic shoe? _____

7. What is the price of the least expensive athletic shoe? _____

8. How many shoes cost less than $80? _____

9. How many athletic shoes are recorded on the stem-and-leaf plot? _____

Prices of Athletic Shoes	
Stem	**Leaf**
6	0 0 2 5 8
7	5 8 9
8	0 5
9	1 8 9 9

Key 6|7 = $67

Data & Graphs Review

Name _____

Use the graph to answer the question.

Most Popular Rides

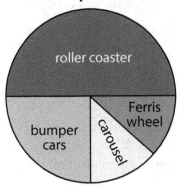

1. Which ride was chosen 50% of the time?
   - ○ roller coaster
   - ○ bumper cars
   - ○ Ferris wheel
   - ○ carousel

2. Which ride was chosen 25% of the time?
   - ○ roller coaster
   - ○ bumper cars
   - ○ Ferris wheel
   - ○ carousel

3. Which rides were the least popular?
   - ○ roller coaster and carousel
   - ○ bumper cars and Ferris wheel
   - ○ Ferris wheel and carousel
   - ○ carousel and bumper cars

Mrs. McCall plotted the number of books her students read during spring break.

```
X
X   X
X   X   X   X
X   X   X   X
1   2   3   4
```
Books Read

4. How many students read three books?
   - ○ 1   ○ 2
   - ○ 3   ○ 4

Test Scores	
Stem	**Leaf**
0	8 9
1	5 5 6 9 9
2	0 0 0

Key 2|0 = 20

5. What was the lowest test score?
   - ○ 0   ○ 8
   - ○ 15   ○ 20

6. How many students scored 20 points?
   - ○ 0   ○ 1
   - ○ 2   ○ 3

7. What score is represented by 1|5?
   - ○ 1   ○ 5
   - ○ $\frac{1}{5}$ ○ 15

Average Temperatures in September

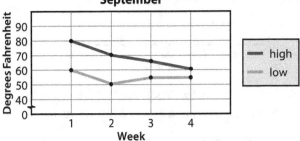

8. Which temperature got increasingly cooler?
   - ○ high temperature
   - ○ low temperature

9. The low temperature increased the most between which weeks?
   - ○ week 1 and week 2
   - ○ week 2 and week 3
   - ○ week 3 and week 4

Use the data to answer the question.

Adam's test scores: 89, 95, 91, 99, 91

10. What is the mean of Adam's test scores?
- ○ 90
- ○ 91
- ○ 92
- ○ 93

11. What is the range of Adam's test scores?
- ○ 0
- ○ 5
- ○ 10
- ○ 15

12. What is the mode of the test scores?
- ○ 89
- ○ 91
- ○ 95
- ○ 99

13. What is the median score?
- ○ 89
- ○ 91
- ○ 95
- ○ 99

Use the table to answer the question.

Grade	K5	1	2	3	4	5	6
Number of Students	20	18	22	15	15	15	17

17. The class sizes range from
- ○ 20 to 17 students.
- ○ 15 to 20 students.
- ○ 15 to 22 students.
- ○ 17 to 22 students.

18. What number of students is the mode?
- ○ 20
- ○ 18
- ○ 17
- ○ 15

19. What number of students is the median?
- ○ 20
- ○ 18
- ○ 17
- ○ 15

Use the graph to answer the question.

14. Did cone sales or dish sales have the greatest increase from May to June?
- ○ cone sales
- ○ dish sales

15. Which had the smaller range, cone sales or dish sales?
- ○ cone sales
- ○ dish sales

16. Would cone sales be expected to have increased or decreased in August?
- ○ increased
- ○ decreased

Use the graph to answer the question.

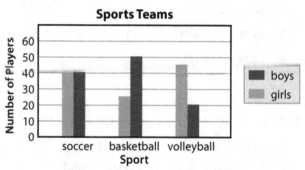

20. What is the double bar graph comparing?
- ○ the number of boys that play soccer
- ○ the number of girls that play soccer
- ○ the number of boys and girls that play each sport
- ○ the number of sports boys and girls play

Mark the answer.

1. Estimate the product of 389 and 517.
   A. 2,000   C. 200,000
   B. 20,000   D. 250,000

2. Estimate the quotient for 3,479 divided into 40 groups.
   A. 60   C. 80
   B. 70   D. 90

3. What is the sum of 13,500 and 27,750?
   A. 30,250   C. 40,250
   B. 35,550   D. 41,250

4. Which expression has a value of 53.9?
   A. $(10 \times 5) + 3.9$ C. 21.3×3
   B. $25 + 25 + 0.9$ D. all of the above

5. Which expression has a value of $5\frac{2}{3}$?
   A. $\frac{10}{3} + \frac{7}{3}$ C. $2\frac{2}{6} + 3\frac{1}{3}$
   B. $5 + \frac{2}{3}$ D. all of the above

6. What is the sum of $\frac{5}{8}$ and $\frac{3}{4}$?
   A. $\frac{2}{3}$ C. $1\frac{3}{8}$
   B. 1 D. $1\frac{1}{2}$

7. Mom bought 10 flats of flowers on Saturday. Each flat had 48 plants. How many plants did she buy?
   A. 0.48   C. 48
   B. 4.8   D. 480

8. Jared had a balance of $255 in his savings account. On June 1 the interest of $1.16 was added to his account. Jared also deposited $65 of his birthday money into his account on the same day. What is his savings account balance now?
   A. $315.16   C. $320.16
   B. $316.16   D. $321.16

9. Marcy purchased a new wallet for $9.95. The tax was $0.60. She paid with a twenty-dollar bill. What was Marcy's change?
   A. $9.45   C. $10.05
   B. $9.55   D. $11.05

10. Janine spent $\frac{1}{2}$ of an hour practicing piano and $\frac{3}{4}$ of an hour doing homework. How much time did Janine spend on these tasks?
    A. 60 minutes   C. $1\frac{1}{2}$ hours
    B. 75 minutes   D. 2 hours

Mark the answer.

11.

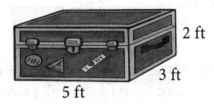

2 ft

3 ft

5 ft

What is the volume of the trunk?

A. 10 ft³   C. 25 ft³

B. 20 ft³   D. 30 ft³

12.

What is the area of the parallelogram if the length is 6 feet and the height is 4 feet?

A. 10 ft²   C. 24 ft²

B. 20 ft²   D. 30 ft²

13.

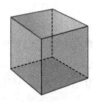

What is the surface area of the square prism if each side measures 2 cm?

A. 10 cm²   C. 24 cm²

B. 20 cm²   D. 30 cm²

14. Which quadrilateral has 4 equal sides and 4 right angles?

A. rhombus   C. square

B. trapezoid   D. rectangle

15.

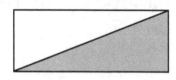

If the area of the rectangle is 28 ft², what is the area of the triangle?

A. 14 ft²   C. 24 ft²

B. 20 ft²   D. 30 ft²

16. Simplify $\frac{24}{5}$.

A. $4\frac{1}{4}$ C. 5

B. $4\frac{4}{5}$ D. $5\frac{4}{5}$

17. Simplify $2\frac{5}{3}$.

A. $2\frac{1}{3}$ C. $3\frac{1}{3}$

B. $2\frac{2}{3}$ D. $3\frac{2}{3}$

18. $(10 \cdot x) + 5 = 55$

A. $x = 3$ C. $x = 10$

B. $x = 5$ D. $x = 15$

19. $\frac{3}{7} = \frac{n}{28}$

A. $n = 10$ C. $n = 24$

B. $n = 12$ D. $n = 30$

20. $\frac{6}{n} = \frac{36}{54}$

A. $n = 7$ C. $n = 9$

B. $n = 8$ D. $n = 10$

21. $\frac{1}{4}$ of 20 = ___?___

A. 4   C. 6

B. 5   D. 7

22. $2.5 \div 4 = $ ___?___

A. 0.5   C. 0.625

B. 0.6   D. 0.65